10643342

DATE DUE

TELEVISION VIEWERS VS. MEDIA SNOBS

TELEVISION VIEWERS VS.

Drawing by C. Barsotti; © 1981
The New Yorker Magazine, Inc.

MEDIA

SNOBS

What TV Does for People

Jib Fowles

THE ELMER E. RASMUSON LIBRARY
UNIVERSITY OF ALASKA

PN
1992.6
F68
1982

𝔰𝔇

STEIN AND DAY/*Publishers*/New York

First published in 1982
Copyright © 1982 by Jib Fowles
All rights reserved
Designed by L. A. Ditizio
Printed in the United States of America
STEIN AND DAY/ *Publishers*
Scarborough House
Briarcliff Manor, N.Y. 10510

Library of Congress Cataloging in Publication Data

Fowles, Jib.
Television viewers vs. media snobs.

Bibliography: p.
Includes index.
1. Television broadcasting—Psychological aspects.
2. Television broadcasting—Social aspects.
3. Television audiences—United States.   I. Title.
PN1992.6.F68   1982   302.2′345   82-40007
ISBN 0-8128-2879-8   AACR2

For my wife, Joy,
and my daughter, Celeste,
with love

Contents

1
On Trial

Ronnie Zamora

"TV Goes on Trial" was the headline of one report in a national magazine, while another demanded, "Did TV Make Him Do It?" A strange, perplexing murder trial in Miami, Florida, was in the spotlight. The trial wore on for ten days, and in the end it all came down to one question: does television have the power to lead a person to commit murder?

The boy who was supposed to be television's dupe was the 15-year-old defendant, Ronnie Zamora. Sitting with his head bowed at the defense's table, the dark-haired youth seemed too frail and scared to be a killer. Everything in his manner showed him to be bashful, not vicious. Yet the crime he had confessed to was particularly cold-blooded and pointless. While burglarizing the home of his 82-year-old neighbor, he had been caught in the act when she returned unexpectedly. In his hand he held her pistol which he had found moments before, and he fired it toward her head. "Oh, Ronnie, you shot me," were her last words, the young burglar had recalled in his confession. She died immediately.

As Ronnie sat meekly by, his attorney, Ellis Rubin, explained to the jury why it was television that ought to bear the blame. "He was suffering from and acted under the influence of prolonged, intense, involuntary, subliminal television intoxication," the lawyer stated. Rubin felt the boy had been undermined by the long hours of violent programs he was accustomed to viewing. "Through the excessive and long-continued use of this intoxicant, a mental condition of insanity was produced." Rubin added, "Pulling the trigger became as common to him as killing a fly."

Television's tyranny over the ninth-grader was portrayed in testimony by his mother, a gentle, heavy-set women with tears welling in her eyes. She depicted her son as a television addict, spending six hours or more every day staring at the screen. "To him, watching TV was the greatest thing in the world," Yolanda Zamora told the jurors. "I used to tell him to turn it off, but there was no way I could stop it." Of all the action/adventure shows he watched, the police show *Kojak* became his favorite. According to Mrs. Zamora, the boy was such a fan of the character played by the bald lead actor, Telly Savalas, that "he even asked his father to shave his head because he wanted my husband to look like him." Completing the defense, Ellis Rubin said that the murder closely paralleled an episode recently shown on *Kojak*.

Rubin did not put his client on the witness stand, but bits and pieces of the boy's life emerged as the trial progressed. As a child he had emigrated with his mother from Costa Rica to New York City. He taught himself to speak English by studying television shows. His mother believed that the boy's attachment to television began when he was placed in front of a set by a baby-sitter and forgotten. The mother remarried, and subsequently the family moved to Miami. The stepfather, who attended the trial intermittently, disliked the boy and was abusive toward him. Ronnie began to watch more and more television, slipping out of bed to view all-night movies.

Rubin called experts to the witness stand who testified that television was fully capable of leading Ronnie astray. A psychiatrist, Dr. Albert Jaslow, said the boy had been improperly influenced by television dramas which "had somewhat blunted his awareness and his capacity to understand his actions." Next came a research psychologist, Dr. Margaret Hanratty Thomas, the author of more than 15 scholarly papers on television violence. She argued that televised hostility and aggression could easily have unbalanced Ronnie.

An extraordinary irony of the trial was that television too was in the courtroom. Like a Greek god watching over the cavorting of mortals, a television camera stood in the shadows at the back of the room, following the proceedings. It was there as part of an experiment authorized by the Florida Supreme Court to see if trials could be telecast without the excess publicity disturbing the course of justice. As the action shifted in the courtroom, the camera panned from the judge across the defense and prosecution tables to the jury box, and back again, recording everything. The presence of television was impossible to ignore completely. During a lull in the trial the curiosity of the jurors got the better of them, and they sent a note to the judge, asking his permission for them to view themselves on the videotape. They would turn the sound off, they allowed, if that would make it all right. The judge was quick to say no, but seeing their faces fall, he volunteered to arrange a private showing for them after the trial was over.

Finally all the defense and prosecution arguments had been laid out for the jury. "If you find Ronnie Zamora guilty, television will be the accessory," attorney Rubin proclaimed in his final statement. He ventured an idea that may have given the jurors something to dwell on: "If you and I can be influenced to buy products by a 30-second commercial, an hour commercial for murder is going to get through to Ronnie's head."

But when the jurors returned to the courtroom, their verdict went against Ronnie. He was found guilty as charged. Although Rubin assured reporters afterward that this unique defense would soon find its place in law, for the moment no one was ready to believe that television had caused a murder. The trial was over, the courtroom emptied, and in a manner of speaking the camera in the back was as free to go as anyone else.

Two Views on Television

In a forceful way the trial of Ronnie Zamora raised once again the question of whether or not television is bad for people. It is a question that continues to haunt Americans, refusing to be settled conclusively. The audience for the nightly telecasts of *Florida vs. Zamora* surpassed Johnny Carson in the ratings, so great was the public's interest in Ellis Rubin's unique defense strategy. Viewers may have been initially attracted to the novelty of a televised trial and the subject of murder, but many stayed tuned because deep in their minds lay the fear that television might truly be harmful. As the account of the trial spread over the wire services to newspapers and television stations around the country, it caught the attention of many more uneasy citizens.

Reasons for being uncomfortable about television are easy enough to come by. The medium is a newcomer in human existence, around for only three decades or so. Yet in that brief time it has forced its way to the front and become Americans' third-ranked activity in terms of hours spent at it. In the span of just one generation, time allocated to television-viewing has climbed past time given to other activities that have been done by humans for tens of thousands of generations—child-rearing, socializing, and game-playing among them. Only the hours for working and sleeping continue to exceed television-watching. The A.C. Nielsen Company, the semi-official scorekeeper of television habits, reports that each American adult now averages about 29 hours weekly in front of a television set. For children the typical amount of time is just one hour less.

Devotion to the medium has reached the point where about half of all kindergarteners, when asked by social scientists, "Which do you like better, TV or Daddy?", will say television. They are picking the force in their young lives which is consistently more available and indulgent. Over half of all American adults look forward so eagerly to their viewing hours that they report making selections well in advance from weekly guides. A letter writer to the *Washington Post* complained, "Recently, I was sitting outside on the front steps of my apartment building when I realized how endemic was the TV disease. All the neighborhood children immediately assumed that I must be locked out."

It's wise to be suspicious about something that is embraced so hastily and so fully. Television could be bad for people despite its widespread acceptance, just as smoking is bad. For all the diversion the shows provide, in the long run they could also have a slow-building corrosive influence. Indisputably the programs have lured people away from the things they used to do before television came along. The hobbies and meetings and conversations that have been shunted aside added a texture to human life that is now disappearing. People don't venture out for entertainment so much anymore because it's easier to lounge at home in front of the set. The situation comedies and action/adventure shows they see transport them away from the real world and cast them adrift. There is some question about how much of this fantasy voyaging a person should have. And as the case of Ronnie Zamora implies, there is also some question about the lessons people take away from these unreal excursions.

Domination by television is suggested by the fact that the majority of Americans now say they get the most and best news from the broadcasters who read it nightly over the channels. Newspapers fell to second place in 1970, Roper polls reveal, and are falling further behind as time passes. This is troubling even to those in the television industry because they know it is print journalism which every day does the real work of gathering information and sorting out the more important stories from the less. Often the first thing a television news director does when he comes to work in the morning is to read the newspapers, to find out what's going on. Many commentators on mass communications have observed that television is the weaker news medium, handicapped in the stories it can carry. There can't be many, compared to newspapers, and what there are should have a visual quality if they are to get on the air. If there is no film footage or no likelihood of an interview, a story may die in the news director's office. So for viewers, being utterly dependent on television news could bring about misperceptions of the world and of what's important.

The other major content of television, the commercial minutes, is even more questionable than the entertainment and news programming. While shows and news are largely appreciated by viewers, the words from sponsors (which account for roughly 15 percent of broadcast time) are just tolerated, surveys relate. People generally feel that the booming advertisements aren't an excessive price to pay for getting what they get from television, but they aren't enthusiastic. Commercials try to make us do things we wouldn't do otherwise, and to the extent they are successful, we need to be wary. No one wants to be gulled by the bewitching images that advertisers assemble and put forth at enormous expense. They invest tens of thousands of dollars in a 30-second spot, and we are left to wonder if that's enough to buy unwanted entry into minds. The worry that Americans are being converted into mindless purchasers, throwing money away for cheap rewards, does not go away.

But the greatest fear, the one least repressible, concerns the effect of television upon children. Adults can probably withstand, but children are impressionable, and the hours they wile away sprawled in front of the set could add up to something harmful and irreversible. In previous decades children were out doing things, building things, creating games and fanciful activities. Now they choose to soak up television's questionable content—hour after hour of cartoon mayhem and situation comedies more sophisticated than need be. There can be something unnerving about hearing a child burst out in a perfect rendition of a television advertising jingle. It can lead any parent to wonder who's in charge here.

It doesn't make people feel any more comfortable about television to learn that the networks think of the audience in callous and economic terms. Television is not so much the business of audience-serving as it is the business of audience-selling. Selling audiences to advertisers is the networks' sole source of revenue. As Les Brown, the one-time television editor of *Variety,* puts it, "People are the merchandise, not the shows. The shows are merely the bait. The consumer, whom the custodians of the medium are pledged to serve, is in fact served up." The viewers' attention, snared by the programming, is sold to advertisers for a multi-billion dollar annual price tag.

It's not surprising that people, having intimations of deception, would think there might be something to Ellis Rubin's version of how Ronnie Zamora became a killer. Ronnie could have resembled a laboratory animal that is administered outsized doses of a seemingly innocuous substance and develops cancer. He might be the true test of the medium, the true measure of its corrupting influence. The case of the fragile boy/man could have been a message about the impact of television in the lives of all Americans.

The jury didn't think so, though. And neither did Russell Baker, the wry columnist for the *New York Times.* When he wrote about the trial several weeks later, Baker endorsed the guilty verdict that overrode the television-causes-murder defense. "Indeed, the belief that television violence breeds social violence is so widely held that to question it seems eccentric. And yet, if people really do tend to become what they see on television, why are working Americans not happier?" Baker asked. That is, if television truly rearranges viewers' behavior (as it is supposed to do in the instance of violence), then why don't people at work in real life act as joyously as people at work in television programs? "In the extratelevision universe one may adventure for weeks before finding a working person who gives outward evidence of enjoying the job. I do not say that the human counterparts of these television workers dislike their jobs, but that they seem to." Baker led his readers to conclude that television is not much good at instruction.

Baker's column hints at a second, less apprehensive point of view regarding television. The real function of television may be not to put lessons in people's brains so much as to take things out. Viewing situation comedies and action/adventure shows may be ways of getting rid of stress and tensions that mount up over the course of a day in the heads of us all. If this were true, it would be a significant benefit for the audience and might explain why people flock to their television sets once the workday is finally over.

This alternative view, which emphasizes the therapeutic services viewers receive from television, can lead to an understanding of the relationship between Ronnie Zamora and television which differs completely from Ellis Rubin's version. Instead of undermining Ronnie, it might be the case that television sustained the youth about as well as he could be sustained, given his situation in life. In his initial estrangement from the mainstream of American life, television provided the child with a helping hand: it was an ever-alert demonstrator of the speaking of English. This would have been the sum of it, as it is for many immigrants, if it had not been for the particular torments of his home. According to the testimony of Ronnie's mother, the stepfather was excessively punitive with the boy. In a stressful household like this, television shows could be a means for vicariously relieving a person's store of resentments. It's not surprising that Ronnie turned to violent programs, and that he saw his stepfather as the aggressive Kojak. But in the end television was not enough to help Ronnie, and the pressures of his home life were converted into an attack on the neighboring adult.

When focus is shifted from the networks and the advertisers to viewers and their needs, as occurs with this second perspective, then even the economics of the system can look different. The most persistent and deceptive myth about television is that it's free to viewers. In truth, the audience pays heavily. Through the purchase of receivers the public has made an investment in the television system which is nine times greater than the investment made by the broadcast industry. Just the annual cost of keeping America's sets in repair equals the after-tax profits of the networks. Some may see these often-overlooked expenses as lamentable, but another way to perceive them is this: Americans pay so grandly, even blindly, to participate in the television system because they receive a great deal in return. It may be that, as viewers, we are buying, and receiving, profoundly important psychological services from the medium.

I should say at the outset that I find this second point of view (the one that draws attention to the therapeutic benefits Americans obtain from television) to be the more reasonable analysis, even though it is less frequently heard. This book exists to present the case that, when all is said and done, television is good for people. As well as examining the favorable evidence regarding television-viewing, I'll also have to explain why Americans have generally elected to misinterpret the significance of the medium, choosing to denigrate it heatedly and needlessly. These highly critical attitudes toward broadcasting I'll be calling Media Snobbery. Why Media Snobs say the things they do, and why they have received such a wide hearing, will be a secondary theme.

The present is a good moment to reappraise what has become our most popular pastime. After decades of hectic growth, television is pausing as if to catch its breath. The potential audience is now saturated, since 98 percent of American households contain a set (4 percent more than have indoor plumbing), and average daily viewing times are leveling off. The next stage in the development of the medium is just beginning, invoking questions of new emphases in programming and new, non-network modes of distribution. Standing between these two periods of change, we should have a good vantage point, if we can only use it wisely.

This probe into the matter of whether television is good for people or bad for people is certainly not the first one ever. It joins a long chain of inquiries, one that stretches back to the very first days of the medium.

2
A Checkered Past

After the Freeze

Easter 1952 fell in mid-April. The weekend, like many Easter weekends on the North American continent, was clear one day and rainy the next. In Saturday's bright, cool weather, warmly dressed Americans stepped out to buy the traditional flowers and plants for their homes. The following day, about the time church services were letting out, showers began to fall along the Atlantic coast, and later in the Far West. Several of the last preseason baseball games were cancelled. Except for what was going to happen the next morning, it was an unremarkable Eastertime.

Early on Monday morning, April 14, an announcement was made in Washington, D.C. which was to reverberate throughout American life. The government agency responsible for overseeing broadcasting, the Federal Communications Commission, had decided that the barriers against the founding of new television stations, firmly in place since 1948, were to be lifted. Now hundreds of broadcasting stations could be added to the 108 already in existence. By the close of the next working day, over 3,000 license application forms had been requested from the Commission's offices. The honor of the first completed application belonged to a group of businessmen from St. Petersburg, Florida, who had worked on it all night and, exuberant and blurry-eyed, had hand delivered it to the Commission on Wednesday.

How the freeze on new stations had come to be imposed in the first place, four years earlier, is a tale not of muscle-flexing by the Federal Communications Commission but of peacemaking. In the years just after the close of World War II, as the young television industry scrambled to supply programming to the American public, the vying companies had gotten into a free-for-all. Sighs of relief were heard in the new network corporate offices when the FCC stepped in on September 29, 1948, and ordered a breathing spell.

Much of the wrangling had been over two issues. The frequencies allocated for television broadcasting had become overly congested too quickly. Stations were recklessly interfering with each other, as if they were trying to blast one another out of the airwaves. Color television was the other problem. The equipment for one proposed system was cumbersome, the reception on the other was poor, and the FCC was in a quandary.

While the FCC hemmed and hawed, the Korean War provided some reason for postponing decisions—scarce resources had to go into war materiel instead of into television equipment, it was said. But steadily the pressures built up behind the gates the Commission had closed. There was little question about the thirst of the potential audience or the desire of the television industry to satisfy it. The number of sets grew 4000 percent in 1948 alone, and during the freeze continued to expand until by the beginning of 1952 there were 15 million in place in American homes. People were purchasing sets long before there were any broadcasting signals they could receive, in fervent anticipation. The television industry was of course aware of this, as it was also aware that in 1951, in spite of the FCC restrictions, stations were showing a profit on their balance sheets for the first time. Economically speaking, the future of the medium looked very bright.

Better produced and more engrossing shows had whetted the audience's appetite during the freeze. Once the Zoomar lens allowed viewers to focus in on the action, sports broadcasting engaged tens of thousands. Comedians began to make use of several cameras and to adapt their material to the new medium. On Tuesday nights the zaniness of Milton Berle's *Texaco Star Theater* brought public life to a standstill. Berle's competition was Bishop Fulton J. Sheen, about whom Berle joked, "We both work for the same boss, Sky Chief." The impertinence of the comic propelled his image into taverns and living rooms situated within range of his broadcasts. As unpolished as these early live shows may have been, they were gobbled up like appetizers by an audience eager for the slow-to-come main course.

By the spring of 1952 Americans' deep fascination with the new technology was compelling the FCC to end its deliberations and arrive at its decisions. Pressures came not only from the public at large, and most sharply from those communities which did not yet have a station licensed for them, but it also came from within the television industry, which looked forward to large advertising revenues and good-sized profits. On the basis of television advertisements Hazel Bishop had seen sales climb from $50,000 in 1950 to $4.5 million in 1952. Other advertisers were clamoring to have the FCC end the freeze and open up new market areas. From all sides the beleaguered agency was being told to get on with it.

Regrettably, the decisions which had been four years in the making were not noticeably wise or beneficial. Subject to the cross-pressures of various interest groups, as well as to a fear of doing something which would have to be subsequently undone, the FCC did not make great strides. The issue of color broadcasts was set aside—a 1950 decision favoring a CBS system was not to be implemented. (Still another decision was going to be made a year later which would endorse the rival RCA system.) As for the crowded broadcasting spectrum, the FCC's answer was to establish new channels operating at ultra high frequency. Many observers at the time thought that the UHF stations would be the poor stepchildren of the television business, and true enough, decades later, only one-half of the UHF stations currently operate in the black. Their average profit of $200,000 in the nation's ten largest markets stands in bleak contrast to the average profit of $3.7 million for channels in the 2 to 13 VHF range.

So wisdom and foresight were not the advantages of the FCC decisions of April 14, 1952. The main achievement was not what the FCC did, but what it stopped doing. It ceased blocking the way, and at that point television could come pouring into American life. To the 108 broadcasting stations of 1952 another 200 were added by 1954, and yet another 150 by 1956. The number of television receivers doubled from 1951 to 1953, reaching 20 million. To look at this cascade another way, only 23 percent of American homes had sets in 1951, and 46 percent did a mere two years later in 1953. By 1955 the percentage had reached the two-thirds mark, and in 1957 it shot by 80 percent. Television was on its way to blanketing the nation.

The great power of the medium was immediately apparent. For nothing more than opening and closing Westinghouse refrigerator doors at the 1952 nominating conventions, a model named Betty Furness became a nationally recognized figure. If the medium could make a person, it could just as easily break him: Adlai Stevenson's windy erudition cost him dearly with the viewing public, which wanted something snappy and visual, like Dwight Eisenhower's one-minute spots. Stevenson, explains media historian Erik Barnouw, "was waging a campaign of the radio age." Television even in the beginning was bringing forth new heroes and new folklore. During 1952 more and more people came to love Lucy and Desi, and when Lucy had her baby on January 19, 1953, 70 percent of all televisions were tuned to the telecast. Her family and the audience's were linked together in a very special way.

When a brand-new technology spreads through a society as rapidly as television did throughout the United States in the 1950s, sometimes an explanation can be found in the fact that a predecessor technology has cleared the way. The first innovation beats down many of the reservations people may have, and the second has a much easier time of it. Radio had gotten Americans used to tubes and dials, and to receiving a daily infusion of news and entertainment in a family setting. Television was an improvement on this, supplying an image with all the rest, but it was not such a radical departure that it called for difficult adjustments. Many of the personalities whom the audience had slowly come to appreciate in radio slid over easily to the new medium—Jack Benny, Amos and Andy, Dave Garroway. (Not everyone from radio made the transition. One who had great difficulty was Fred Allen, who snarled in 1952, "Television is a triumph of equipment over people, and the minds that control it are so small that you could put them in the navel of a flea and still have enough room beside them for a vice-president's heart.") Because radio had predisposed Americans to electronic communication, television was accepted into living rooms at a rate twice as fast.

There is more to it than that, though. Radio helped, but the root causes of the love-at-first-sight romance with television lay much deeper.

The 1950s were odd and trying times, very different from other periods in the nation's history. During the two preceding bone-jarring decades Americans had been struck first by an economic depression whose gravity was beyond comprehension, and then on the heels of that by a global conflict whose butchery and scale were without precedent. The savagery of the '30s and '40s was weathered only by marshalling every ounce of fortitude and energy, and when it was all over, Americans were done in. Those events had called into question every principle of life as it was known. They had made people skeptical of the economic foundations they had to depend on and skeptical of the essential nature of their fellow human beings as well. Nothing looked quite the same when the turmoil finally subsided.

The American public resembled a person who had been through a terrifying automobile accident, who had behaved conscientiously in the aftermath, and who only some time later began to shake and tremble. The 1950s were the time for quaking. Demons were sensed everywhere, menacing from without, destroying the body politic from within. These forces of evil, in the form of International Communism, would have to be exorcised. The job fell to the junior Senator from Wisconsin, Joseph McCarthy. And around the world, the United States mustered its remaining strength and took its stand. The Cold War was on.

Tired, frazzled, Americans wanted nothing so much as to retreat to the security of their families and the isolated comfort of their living rooms. So they bred children and bought television sets, and thus fortified themselves. "In the early 1950s, by the millions," says Jeff Greenfield, media-watcher and the author of *Television: The First Fifty Years,* "we took television into our homes and closed the doors behind us."

The audience of the '50s had very exacting demands of the newcomer who was to share their shelter. Tolerance of ambiguity was at a low ebb: things on television had to be this or that, right or wrong. Evil was to be unmitigated and good was to be triumphant. Since intrigue and war shows cut a little too close to the quick, the Western provided a setting of proper distance where these forces would work themselves to a suitable conclusion. Westerns began to take hold in 1955, and by 1959 there were more than thirty of them regularly scheduled each week. Americans also wanted shows that shored up their revitalized and, with the baby boom, repopulated family life. These shows would have to idealize the home and stimulate warm feelings for it. The television industry was quick to oblige with a flurry of "Father" shows. Not only adults needed new models—the young did even more so. When Dick Clark's *American Bandstand* began to be televised nationally in 1957, teenagers saw not just how to dance, but also how to dress, how to behave, what to emulate. And not the least of it, the very young needed orientation— every American boy wore his Davy Crockett coonskin cap in 1955, the year that over $300 million of Davy Crockett merchandise was sold through the efforts of Walt Disney.

It had taken a few steps for television to become such an intimate of Americans. Like any courtship, this one had extended over several different settings, each one increasingly private. First there was the meeting at the neighborhood tavern. Here the new device could be scrutinized with some degree of anonymity, in the company of mere acquaintances. If it wasn't to prove out, not much would be lost. The bars, with business slipping away because of the post-World War II retreat into homes, had risked investing in the new gadgets to lure their customers back. The irony was that many stayed only long enough to size up the picture tube, then left and never returned. They had purchased one for themselves.

Once in the home, television went through the ritual of being introduced to friends and relations. The sets' owners would invite people over for a television party, where popcorn would be served and conversation would be halted when the show came on. "Well, what do you think of it?" they asked each other. They thought well enough of it to stop asking anyone over anymore. They did not want to dilute the pleasure it gave them. So the marriage was made and the honeymoon was under way. Now it was just the individual—sometimes a family cluster of individuals—and his set. By the middle of the decade typical viewers had their sets on five hours a day.

In any sort of union, and especially for a heartfelt one like this, some sort of accommodation has to be made by both parties. It would be strange if a force so omnipresent and riveting did not cause some changes in household living patterns. We had to move over a bit if we were going to share our quarters with television.

Living rooms were rearranged. The set was placed where it could be seen from the largest number of seats, or at least where the heads of the household could have the best view of it. Other chairs and furniture were repositioned out of the line of sight. There might be some grumbling from someone who did not want a favorite chair moved, but quickly things settled down and the set became a part of the room. It was further knit into the household by the bric-a-brac and photographs that a housewife might crown it with.

More fundamentally, the use of time was rearranged. In order to spend five hours a day watching television, lives were rescheduled. Some observers feared that viewing time was won by giving up worthwhile activities, but it does not seem to have been so. People still read books and went to church. The time found for television came primarily from elsewhere. Viewers stayed up an hour or two later in the evening. And they redirected their attention from other fairly low-grade diversions. That is, they stopped going to minor league baseball games and Grade B movies, stopped reading fan magazines and comic books, stopped listening to radio serials. According to Martin Mayer, the author of several articles and books on television, "Television battered its rivals out of shape."

By the close of the '50s television was woven deep into the fabric of American life. It was a presence as great as that of another spouse. And perhaps better—it was invariably bright, alert, ready to befriend the viewer. Just as in earlier times when people had brought the sink and then the commode into their residences, here was one more piece of essential equipment. By 1960 virtually all Americans were watching, no matter what they claimed. It was a national mania of a magnitude never experienced before.

The '50s had also brought changes in the content of television. And wasn't it too bad, critics began to say aloud, that the changes weren't for the better? At the start there had been those well-wrought, dramatic productions, live from New York. Something over 300 hour-long plays had been written and produced before 1955. The most famous of all, Paddy Chayefsky's *Marty,* was aired on May 24, 1953, and the lead actor, Rod Steiger, could scarcely believe the response. "People from all over the country and all different walks of life, from different races and religions and creeds, sent me letters. The immense power of that medium!" But the enthusiasm for quality drama began to slacken soon after. Some analysts say it was because the audience was changing: the more affluent who had purchased sets in the early '50s had a taste for theater, but the diffusion of television as the years went by meant a dilution in the quality of the shows. Other observers believe that the end of televised drama resulted from developments within the industry. For a variety of technical and economic reasons, production had to shift from live to filmed. By using film, shows could be more effectively edited and produced, and they could also be resold, or "syndicated." The shift from live production, with its intimacy and occasional gaffes, entailed a shift away from the sound stages of New York City, and from the traditions of American theater. Transplanted to Hollywood by the end of the '50s, television production was undertaken by less theatrically conscious writers and crews.

In addition, control of the shows was switching from advertisers to networks. The Philco and Goodyear series became things of the past; now the network would put a show together and sell advertising time on it. In part this change was the outcome of rising production costs, which were exceeding the pocketbook of any single advertiser, and in part it resulted from the growing conviction that sponsors could make the best use of their advertising budget if they could spread it around more, buying time on different shows and different stations. So now there was less interest in associating a trademark with a prestigious show, and more interest in reaching the largest possible audience, no matter how tawdry the offering. Rip-'em-up action and adventure became the hallmark of network television by 1961, when the highest ratings went to ABC's *The Untouchables,* a show awash in gore and mayhem.

No matter how low the content sank the audience still watched. It made little difference what was on, detractors of television stated. Inertia, not standards, dictated viewing. In the beginning of the decade the choices of the audience seemed to exert control over the medium, they said, but by the end it appeared to be the other way around, in that television had some kind of hypnotic control over viewers.

It was not just that adults were being mesmerized. That would be bad enough. But children were too. All the babies produced in the decade of the revived family were sitting in front of the cathode-ray tube just as fixed-eyed as their parents. Typically a child would absorb a three-hour dose of television, day in and day out. Critics asked if this was the best thing, to expose our tenderest flowers to such stuff? Levels of discomfort climbed.

As people began to squirm and to raise questions like this, the commercial aspects of television were becoming more visible. Eighty-five percent of all homes could be penetrated by the 500 broadcasting stations in existence in 1957, to the joy of sponsors. The $450 million dollars they invested in television advertising in 1952 tripled by 1956 and continued to skyrocket. Every company with something to sell and enough money to do it raced to buy time on the network shows. Political parties spent $6 million on television for the 1952 presidential election, and then almost doubled the figure four years later. A successful show for a sponsor amounted to a money-minting machine. Hazel Bishop, the leading manufacturer of cosmetics by virtue of its television advertising, would have remained so throughout the decade if Revlon had not come along and beat the company at its own game. Revlon paid for the enormously popular *$64,000 Question,* which captured 85 percent of the viewing audience in 1955. The board chairman of Hazel Bishop, Inc. was called upon to explain to his stockholders in January 1956 why a surprising drop in profits had occurred the previous year. It was due, he glumly confessed, "to a new television program sponsored by your company's principal competitor which captured the imagination of the public."

It all backfired, of course. Wealthier and humbler for the experience, Charles Van Doren blew the whistle in 1959. The Columbia University English tutor confessed that the quiz show he had starred in had been rigged, and he had been fed the answers. "I have deceived my friends," he lamented in Congressional testimony, "and I had millions of them." Each one of those millions of friends was left with a bad taste in his mouth. President Eisenhower said sternly that the quiz show deceptions were "a terrible thing to do to the American people." They had made the audience even more uneasy about the new medium. Were viewers becoming dependent on something that wasn't worth their trust? If television was deceiving us on this count, was it also deceiving us on others?

The Steiner Poll, and the Dodd Hearings

Some of the public's apprehension about television was revealed in a national survey conducted in 1960 and paid for by the Columbia Broadcasting System. In March and April a sample of 2,500 viewers was contacted by two polling organizations, the National Opinion Research Center of the University of Chicago and Elmo Roper and Associates. According to Dr. Gary Steiner, who ran the study from the Bureau of Applied Social Research at Columbia University, television received generally good grades from the public. When asked, "How do you feel about TV in general?" favorable responses outran unfavorable ones about 2½ to 1. "Relaxing" and "interesting" were the adjectives people mentioned first. Half of the respondents volunteered that television made them feel "satisfied and peaceful." Especially when compared to other inventions and other media, the appreciation of the public for television was clear.

Nevertheless, whether or not the Steiner study was looking for reservations about television, it did uncover them. Parents were ambivalent about children watching the set. About 4 parents in 10 said they had rules regarding what their offspring could watch, and nearly 3 mothers in 10 said that they personally knew of an instance in which a child was harmed or did something harmful as a result of watching television programs. The respondents were also ambivalent about their own viewing. In order to elicit deep-lying feelings and attitudes about television, the interviewer would display a sketch of people watching a show, and then ask if the people in the picture might be thinking some of the sentiments which the interviewer proceeded to read. When the interviewer voiced the line, "Am I lazy!" about half of the 2,500 respondents agreed that was a thought in the minds of the depicted television viewers. Thirty-one percent thought the sentence "I really should be doing something else" applied, and 20 percent concurred with "Another evening shot." A full 15 percent subscribed to "I'm a little ashamed of myself for spending my time like this."

Americans needed help in understanding their entanglement with television, and it is not surprising that it would be a politician, attuned by personality and profession to the temperament of the public, who would most fully respond to the need. Senator Thomas J. Dodd of Connecticut was a debonair and photogenic man of considerable ambition. In his youth he had toyed with the idea of becoming an actor, and in later years he was never far out of touch with his constituency or his audience. Acting as chairman of the Senate subcommittee on juvenile delinquency, Dodd opened hearings on television violence in June 1961. It was not a new concern—Senator Estes Kefauver had raised the same question ten years earlier—but it had become more pressing. The public wanted answers this time.

At the outset there were unnerving disclosures regarding the networks' preoccupation with violence. *The Untouchables* and ABC received a particularly tough time of it as the result of in-house memos which had fallen into the investigators' hands. Some people felt an irrepressible pleasure in seeing executives from the three networks twist and turn in the witness chair.

But the pace of the hearings became sluggish when Dodd called on social scientists to give evidence about the damage televised violence was supposed to be inflicting. Some said one thing and some said another. The highly respected communications scholar Wilbur Schramm spoke on the basis of his own extensive research projects and said that only a child psychopath could be triggered to a violent act by the sight of conflict on television.

On the other side of the question was the now famous Bobo doll. A Bobo doll is an inflated figure several feet high which is weighted on the bottom so that it will obligingly roll back to an upright position if knocked down. A research scientist named Albert Bandura had made an eight-minute film of a lady whacking a Bobo doll with a mallet while gleefully shouting "Pow!" and "Sockeroo!" Bandura went on to demonstrate that 3-to-5-year-old children will imitate the lady if they are given the chance immediately after seeing the film. Apparently the conclusion was that, for the greater majority of children, violence in the media will instigate violence in real life. ("Apparently" because some time later Bandura himself shied away from this, saying, "Some people have misunderstood the purpose of my Bobo doll experiments. They were designed only to determine to what extent a child *learns* about behavior from television. One would use quite a different experimental procedure if one were testing for aggression or punitive behavior *to other people.*")

It all ended inconclusively. Almost a year later, in March 1962, Dodd wrote the Secretary of Health, Education, and Welfare and suggested that the Secretary take it upon himself to contact universities across the country and ask them to do more research on the problem. That was the sole product of two years of probing.

Erik Barnouw, writing in *Tube of Plenty,* thinks that the hearings might have ended on such a false note because Dodd had been compromised by the networks. It is true that Dodd was increasingly seen in the company of network higher-ups, and that at this time he suddenly started to live on a grander scale than before, turning his Connecticut retreat into a lavish estate. He went so far as to appoint the son of a media executive to his subcommittee staff. Edith Efron, on the other hand, feels that the hearings ended in indecision because of the intrinsic difficulty of the problem—at the time no one could say with any certainty what the results of viewing television violence might be. "The entire Dodd hearings were nothing but one vast deadlock over the central issue of children's comprehension, which was chewed over and over in a variety of forms," recalls the writer for *TV Guide.*

In either case, the American public was shortchanged. Confusion was even greater than before the hearings had begun.

Marshall McLuhan

At the time that his singular and widely-read treatise *Understanding Media* was published in 1964, Marshall McLuhan was a professor of English literature at Canada's University of Toronto. The book was no common scholarly work brimming with footnotes and precise, hairsplitting discourse. The style of presentation was a unique one, relying on strange aphorisms ("The medium is the message") and unexpected terms ("hot" versus "cool" media). The initial readers recognized they were turning pages they had never seen the likes of before.

But still they read on. They were searching determinedly for what they thought the title promised—an explanation of the ways the media were rocking their lives. To get this information they were willing to endure some of the most peculiar sentences ever set down by any author. "Any invention or technology is an extension or self-amputation of our physical bodies, and such extension also demands new ratios or new equilibriums among the other organs and extensions of the body," the book related, and at another point, "The electric light is pure information." In spite of these baffling notions the first readers were soon joined by hundreds more, and then thousands more, until hundreds of thousands of people had been exposed to what some detractors would say was McLuhanacy. Even arch-critics like Sidney Finkelstein have felt called upon to explain why, for all its supposed shortcomings, *Understanding Media* has been so popular: "A basic reason for the appeal of McLuhan's book is its central subject of television. An increasing number of Americans feel that television, far from being simply another form of entertainment which they can take or leave at will, has insinuated itself into their lives and is affecting their minds with the ineradicability of a drug habit." Nagging uncertainty about television—that was the reason that McLuhan was so much in demand. He was the first intellectual to concede that television was just as momentous and mysterious in its ultimate effects as many members of the public had intuited.

What was McLuhan saying about the coming of television to the diligent reader of *Understanding Media*? Confusing things, conflicting things. It was difficult to learn if he perceived it as good or bad. Most authors who subsequently challenged McLuhan's theories on the media did so in the belief that McLuhan endorsed the changes he saw accompanying television. After all, wasn't a positive view being suggested when he wrote, "In the electric age we wear all mankind as our skin"? And didn't he say that for the individual this integration of humanity carried with it the opportunity for "uniqueness and diversity that can be fostered under electric conditions as never before"? But McLuhan also made less cheerful observations. "The new media and technologies by which we amplify and extend ourselves," he lamented, "constitute huge collective surgery carried out on the social body with complete disregard for antiseptics." The patient may be going under: "As a cool medium TV has, some feel, introduced a kind of *rigor mortis* into the body politic." McLuhan went on to say, "Thus the age of anxiety and of electric media is also the age of the unconscious and of apathy." As well as apathy, McLuhan also linked television with "myopia," "hallucination," and "suicidal autoamputation." (Five years later, in 1969, McLuhan confessed his true feelings to interviewer Eric Norden when saying, "I don't like to tell people what I think is good or bad about the social and psychic changes caused by new media, but if you insist on pinning me down about my own subjective reactions as I observe the reprimitivation of our culture, I would have to say that I view such upheavals with total personal dislike and dissatisfaction.")

If it was hard to understand whether McLuhan thought good or bad things were generally happening, it was harder still to learn exactly what he believed television was doing to people. The key to it seemed to be the way a medium such as television was absorbed through the human senses. In McLuhan's theory, because the picture on the television screen lacked high definition, "the TV image requires each instant that we 'close' the spaces in the mesh by a convulsive sensuous participation that is profoundly kinetic and tactile." Kinetic and tactile viewing was not an easy concept to comprehend, although it was good to know that it induced participation. Increases in audience participation were the earmark of what McLuhan termed a "cool" medium: "A cool medium leaves much more for the listener or user to do than a hot medium. If the medium is of low intensity, the participation is high. Perhaps this is why lovers mumble so." The upshot of it all was that pre-television patterns and sequences of life were being wrecked. Speculated McLuhan, "Just as TV, the mosaic mesh, does not foster perspective in art, it does not foster lineality in living. Gone are the stag line, the party line, the receiving line, and the pencil line from the backs of nylons."

McLuhan's unique ideas about media effects were grounded in his own special version of history. In the same way that modern-day trans-formations should be credited to television, he argued, historical changes resulted from earlier media innovations. Specifically, the creation of the alphabet destroyed oral communication and started mankind on the way to fragmentation and small-mindedness, and the invention of the printing press completed the process. No matter that most scholars of world history would say that the two turning points in mankind's evolution were the start of agriculture and the onset of industrialization, for Marshall McLuhan it was the alphabet and, more outstandingly, the printing press. "The hotting-up of the medium of writing to repeatable print intensity led to nationalism and the religious wars of the sixteenth century," he proposed. It also led to social fission, specialization, linear-ity, and other such effects. It did this, to McLuhan's way of thinking, by biasing our senses toward the visual and our minds toward the sequential.

Critical response to *Understanding Media* was particularly severe on these points. The visual orientation of human beings extends not just back to the 15th-century printing press, but back some 60 million years to our ancestor species who began to negotiate through the treetops. Those animals needed a good sense of sight if they were to jump from branch to branch without mishap. From that species down through the eons to *homo sapiens*, vision accounted for the greater proportion of each individual's incoming stimuli. The alphabet and the printing press did not create calculating populations—they were the products of man-kind's age-old quest for more vision, more stimuli, more information.

If McLuhan was misinformed on those counts, was he also wrong about where television was steering us? He foresaw the emergence of a "global village" in which we will come to be "retribalized" and to share in the feelings of community last experienced by primitive men. Leaving aside questions about the real nature of tribal life, this is an unlikely scenario for the future of industrial society, where life is structured, for better or worse, along lines of competition between corporations and between nations. McLuhan prophesied a world operating without eco-nomics or politics, when most others who ponder the future anticipate nothing of the kind.

All in all, the value of McLuhan's speculations on the significance of television remains questionable. McLuhan himself did not stand behind them. In the same 1969 interview in which he expressed his true feelings about television, he also remarked, "As an investigator, I have no fixed point of view, no commitment to any theory—my own or anyone else's. As a matter of fact, I'm completely ready to junk any statement I have ever made about my subject."

When read or listened to, McLuhan was often received with resentment. This must have puzzled him, since in his mind all he was doing was conducting an innocent, academic activity. But for his audience the inquiry was far less casual. The reader of *Understanding Media* was put in the position of the person who, getting on a strange bus and not knowing where his stop was, asked another rider, who replied, "Watch me and get off one stop before I do." People did not know how far to accompany McLuhan on his explorations, and they ended up being frustrated and indignant.

McLuhan articulated deeply-felt questions about the effects of television, but his replies may have been so personal and confounding that the final result was to make better answers even harder to find. If Americans were reeling from the way television had burst into their lives in the 1950s, McLuhan had sent them spinning even faster.

Bower's Survey, and the Report to the Surgeon General

There are two things that can be said incontrovertibly about television for the period between the publication of McLuhan's *Understanding Media* and the present: people have continued to view it, and they have continued to be perplexed about its ultimate effects. These two themes were apparent when the 1960 Steiner public opinion survey regarding television was updated in 1970. Also paid for by CBS, the second survey was conducted by Robert T. Bower, past president of the American Association for Public Opinion Research and the National Council on Public Polls. The national sample of 2,000 adults told Bower they were watching television more than ever. When asked about the proportion of programs "you generally watch" judged to be "extremely enjoyable," the average of 50 percent was six percentage points higher than ten years before. But when queried about their satisfaction with the medium overall, the figure slid by 13 percent. As one measure of the public's wariness, the proportion of parents with definite rules about their children's viewing had increased 3 percent over the decade.

An intriguing feature of Bower's 1970 survey was his construction of what he called an "equal opportunity" measure of viewing hours, which deliberately handicapped those who had the unfair advantage of being at home during the day. This ingenious tabulation revealed that, no matter what their background or attitudes, people all watch about the same amount. It documented, Bower said, "a very even spread of viewing among old and young, males and females, college and grade school educated, blacks and whites, wealthy and poor, parents and non-parents, and viewers of different religions, ideologies, and political persuasions." Very few escape the pull of the medium. Bower found that television avoiders made up about 5 or 6 percent of the population, and that there was no discernible social pattern to them. They were not clustered within any of the various social levels or groups, a fact which suggests that abstention from television is not related to education or age or anything other than the psychology of a rare individual. As for the rest of us, we are all hooked.

The opportunity to reach final answers about our uneasy addiction seemed to present itself in 1970 when Senator John O. Pastore, Democrat of Rhode Island, charged the office of the United States Surgeon General with determining what negative effects television might be having on the population. The social upheavals of the time and the desire to know their causes stimulated interest in the study, and it was generously funded. Dozens of experiments and research efforts were commissioned to scrutinize the core issue: does televised violence have a harmful effect upon children?

The investigation by the Surgeon General's Scientific Advisory Committee on Television and Social Behavior concluded in a less decisive fashion than it began. The *New York Times* obtained a copy of the final report prematurely and ran a front-page story headlined, "TV Violence Held Unharmful to Youth." No, no, people connected with the study protested, that was not to be the conclusion. When it was officially released on January 17, 1972, Surgeon General Jesse Steinfeld stated, "The experimental findings are weak, and are not wholly consistent from one study to another. Nevertheless, they provide the suggestive evidence in favor of the interpretation that viewing violence on TV is conducive to an increase in aggressive behavior, although it must be emphasized that the causal sequence is very likely applicable only to some children who are predisposed in this direction, and TV is only one of the many factors which in time may precede aggressive behavior."

The million-dollar budget had produced a high degree of equivocation. The causative link between televised violence and aggression in real life was not firm; the number of children who were actually susceptible to this influence was not even guessed at—it could be a handful or a legion; the role of inciting agents other than television—such as the type of family life—was not examined. Interspersed through the report are the scientists' complicated caveats and disclaimers which Edith Efron paraphrases, "If we translate all this into basic English, it reads: We don't understand children. . . . Every child in America is different. . . . TV violence is just one of a billion stimuli impinging upon him. . . . We don't know how to test its effects scientifically. . . . And when we try we usually don't understand our own findings." Dr. Eli A. Rubinstein, vice-chairman and research director of the Advisory Committee, conceded, "We have barely scratched the surface of understanding this complex phenomenon." And Dr. George Comstock, senior research coordinator of the project, admitted, "The central question of the role of television violence in aggressiveness cannot be taken as fully resolved. The case is not beyond reversal." When the scientists who had actually conducted the studies were polled, the fence-sitting was unquestionable. Asked by Professor Matilda Paisley of Stanford if they believed that television violence causes aggression, about half said no, they didn't.

Senator Pastore accepted the report, complained, "It looked to me like a little double-talk," and convened hearings. His colleague, Representative John A. Murphy of New York, who was the first witness, claimed the report was inconclusive because the ruling 12-man Advisory Committee was loaded with people who had ties to the television industry. The Surgeon General, perhaps goaded by this and other criticisms, testified, "It is clear to me that the causal relationship between television violence and antisocial behavior is sufficient to warrant appropriate and immediate remedial action." But few seemed truly convinced. It is the final comment on the report that "appropriate and immediate remedial action" was never taken. Just as with Dodd's hearings ten years earlier, no policies resulted and no laws were written. Except to add to the pressure on the networks to cut back their violence offerings, the whole project might just as well not have been undertaken, for all the official influence it had.

In the aftermath of the 1972 report to the Surgeon General, public misgivings may have swelled, but not to the point where dedication to television has ever been threatened. The truth of the matter is that people can't seem to live without it. A poll taken in 1975 asked, "Not including your family, what do you consider the three most important things that you now have in your home?" and television topped the list, mentioned more often than refrigerators, stoves, beds, pets, and all the rest of it. When the *Detroit Free Press* tried to find families to give up television for a month in 1977, in return for $500, they had to approach 120 families before they got their five volunteers. "I was surprised that there really did seem to be an addiction to television," later commented reporter Cathy Trost about the five families. "Some of those people almost literally went crazy. They didn't know how to cope." One husband barricaded himself behind his daily newspaper and stopped talking to his wife. "I think he's suffering from withdrawal," his wife sighed. Two people in the study began to chain-smoke. One woman became so nervous she took tranquilizers and talked with the cat. "It's so quiet," she said. "I talk to the cat, and he looks at me like I'm crazy." At times almost everyone involved became depressed, bored, or nervous. When the sets were finally restored a month later one wife gushed about television, "It's just a nice warm feeling that creeps over you."

Equivocation

The passing years have brought us to the point where one-half of all Americans—those born since the 1948–52 freeze on new television stations—have never lived in a world without antennas and interference, channels and shows. Television has become as much a part of existence as the rise of the sun and the draw of gravity. The number of sets in operation now amounts to one for every other person, and according to the Nielsen rating organization, each one of the 116 million sets is turned on for over six hours a day. To fulfill the high demand, the volume of television programming has reached five million hours annually. The question becomes more insistent—is our involvement with television a dangerous one?

Over the history of the medium, all the major efforts to untangle television's effects have ended up equivocating. Each has produced new facts and insights, and none has led to a fully lucid understanding. Polls have told of the ambivalence of the public; hearings have presented experts on both sides; massive research projects cannot find the evidence for clear-cut conclusions; and the probes of a genius have clouded rather than illuminated the situation. The two opposing perspectives on television—that television is bad for people and that television is good for people—have remained badly intermeshed in each of these inquiries.

The reason for the continued equivocation over television's influence, I believe, is that the individuals involved—poll-takers and experimenters, respondents and subjects—have not been able to take leave of condescending, negative biases against the medium. They could not look upon television with total objectivity because dearly held beliefs would have been violated if they had. Convinced that television is somehow contemptible, yet operating with an awareness of the extensive and pleasurable viewing done by themselves and others, their reports have taken on a vacillating tone. The culprit here is Media Snobbery, I think we'll come to see. The highbrow, reproachful attitudes of Media Snobs can make it difficult to think clearly about television or appreciate why 100 million Americans will switch on their sets in the course of an evening.

3

What We Truly Want (and Get) from Television

Six Clues

When a person sits down in front of a television set, a transaction is going on, although the individual isn't likely to think of it that way. The viewer is tacitly agreeing to swallow a few commercials and consider some advertised products. In return for this attention (which turns out to be fractional), the viewer wants to be sent certain kinds of content— phantasms of a very special sort. If they aren't televised, the viewer won't bother to watch.

What is it exactly that human beings want from the medium? Some observations about viewing habits provide clues.

1. *Television-viewing is a personal, private activity.* It wasn't this way originally, at least not to all appearances. When television first came in it was a family activity. Parents and children would arrange themselves before the set, and program choices would be made consensually. But as receiver costs plummeted and average family size shrank, the number of viewers per set continued to drop. Television historian Leo Bogart comments, "The most important change taking place in television is its transformation from a medium of generalized family entertainment into one of more intimate, personal viewing."

The diminishing number of viewers at each set is a visible sign of a psychological reality—that viewing is not a social activity like softball or conversation, but rather the private engagement of one mind by a program. Television is something the individual does for himself alone.

2. *Television-viewing is an enjoyable activity.* Respondents in all national surveys on attitudes toward television testify to the pleasure the programs give them—often to the amazement of the person reporting this. "In view of the frequent and heavy criticism leveled at television, it is somewhat surprising that rather positive attitudes toward specific and general programming are prevalent," conceded a primary researcher for the 1972 Report to the Surgeon General. Eighty-two percent of the programs watched evoked a pleasurable response from the researcher's sample of viewers.

3. *Television-viewing is a needed activity.* Newspapers occasionally carry stories about people who go berserk when someone interferes with their television time. Charles Green of East Palo Alto, California, stabbed his sister to death with a hunting knife after she took out the electrical fuses so he would stop viewing. In Lawtell, Louisiana, John Gallien shot his sister-in-law because she kept turning down the volume. Gruesome tales like these serve as reminders of how important television can be to people.

One way to gauge the need for television is to remove it from households and study the extent of the deprivation. But as the reporters who did the series on the five television-less families for the *Detroit Free Press* learned, it is difficult to find volunteers willing to surrender their set for several weeks. Even one week was too much for most people to risk, mass communications professor Alexis Tan discovered before managing to assemble a sample of 51 adults. In a paper published in the *Journal of Broadcasting,* Tan reported how the subjects fared once the plugs to their sets had been taped up so they couldn't use them. The conclusion:

"TV is a very important force in many people's lives. This was indicated in the present study by the difficulty the researchers had in recruiting participants, by ratings given TV as a primary source of information about current events, news analysis, and entertainment, and by how some respondents said they had become disoriented without the medium." When asked which medium they would give up if they had to give up one, 49 percent said magazines, followed by radio and newspapers tied at 22 percent each, while only 8 percent of Tan's sample would relinquish their television sets.

4. *Television-viewing is a casual activity.* An incongruity. People need television, but their viewing is done offhandedly. This seeming contradiction will prove to be informative about the real function of the medium.

In the early 1960s researcher C.L. Allen persuaded 95 households, with a total of 358 individuals, to accept time-lapse movie cameras in their living rooms. At four frames per minute the cameras monitored the television sets and viewing areas. What Allen learned upon analyzing the films was that for 20 percent of the time the sets were on, no one was in the room. For another 20 percent of the time, the people in the room were not looking at the screen. Inattention was greatest when commercials were running, approaching 50 percent of the time.

A similar study, with similar findings, was done in 1972 by Robert Bechtel for the Report to the Surgeon General. Into 20 homes Bechtel and his associates put video cameras which were activated whenever the set was turned on. Viewers' attention was found to rise according to what was being telecast, ranging from half for the commercial time up to three-quarters for movies. Bechtel's deciphering of the videotape footage resulted in a long list of what his subjects were doing when they were not looking at the screen, which included untying knots, posing, crawling, scratching someone else, picking one's nose, undressing, singing, pacing, and so on.

It doesn't take cameras in living rooms to learn that viewer nonchalance is high—people admit it freely when asked. They confess they are frequently doing something else when they are watching, and that about a third of their program selection is hardly selection at all, but simple acquiescence. They kept what the channel was carrying when the set was switched on. Once they began looking at a show their attention often wandered; a third of their programs, they estimate, are not viewed to the end. What they do view they generally can't recall the following day.

5. *Television-viewing is mostly an evening activity.* This is so plain it is easy to overlook. Viewing is twice as heavy during the prime-time hours than during the daytime. By midevening about 40 percent of the American population will have their sets on.

To say this another way, most television viewing occurs after the day's work is done and while people are edging toward sleep.

6. *Television-viewing has come to displace principally a) other diversions, b) socializing, and c) sleep.* These are the three types of activities hurt most as television diffused. They may have suffered because whatever they did for people, television did it better. This suggests that television performs a function that overlaps them all.

Time budgets are the way this displacement has been documented. These are measures made by social scientists of how Americans allocate their hours, and of how the allocations have changed over the decades. To create time budgets a sample of a population is asked to keep diaries for 24 hours regarding how they spend their time, and then all the diaries are averaged to obtain national figures. Much of the research along these lines has been conducted by Professor John A. Robinson, first at the Survey Research Center of the University of Michigan and then at the Communications Research Center, Cleveland State University.

Robinson's major study was carried out in 1966 with a sample of 2,000 American adults. When these data were compared to time budgets from the 1930s, the greatest change observed was in television-viewing time, needless to say. "Thus, at least in the temporal sense, television appears to have had a greater influence on the structure of daily life than any other innovation in this century," Robinson wrote.

Time for television was drawn from what Robinson called its "functional equivalents"—listening to the radio, going to movies, reading magazines, attending sports events. These diversions fell off, and so did the amount of time spent fraternizing. In a 1975 update Robinson released new data showing that the hours spent in social interaction by Americans had continued to drop as time with television increased.

Television has been replacing the easeful activities which help people relax perhaps because it serves better. How television accomplishes what banter and other diversions used to do is implied by the third activity it has chiselled away at: sleep. Robinson reflected provocatively about the television set owners in his time budget studies, "Particularly interesting, they obtained an average of almost 15 minutes a night less sleep, suggesting an interesting trade-off or functional relation between television and sleep."

These six items are clues to the mystery of what viewers want and are getting from television.

Fantasies

The dreams of a sleeper and the programs of a viewer have a number of striking similarities. Both dream content and television content consist of a highly disjointed procession of images and doings. Television programming can shift from a show to commercials and back to the show again, then move to a station break and news summary as well as several more advertisements before a new show comes on. There is little that is orderly about the sequence of material. A novice viewer of television finds himself in a fitful, inchoate world. Additionally, both dreams and television are chiefly visual in their impact, with vivid settings and characters. The abstract has been left trailing behind, while the sensory has come to the fore. These images on the screen and in dreams are not one-shot creations; they are symbolic, resonating with patterns locked deep in brains.

Both dreams and television make extensive use of wish-fulfilling material. Things happen there which could never transpire in real life. Adventures and romances, loyalties and successes, are all too outsized to be duplicated in the everyday world, as much as people might long for them. Dreams and television are where desires get obliged. And yet, despite the richness of the two, and the poignancy they can have for humans, both dreams and television are difficult to recall once experienced. With some exceptions they don't seem to have a place in conscious recollections.

What television shows and dreams have in common is that both are fantasies. The items in the list of shared qualities—illogical, visual, symbolic, wish-fulfilling—apply to any fantasy. Fantasies are products of the human imagination, sensed visually, and felt to be emotionally engaging. When fantasizing, people experience characters, action, sequence, and at the end are often left with deep if fleeting sensations. If we are going to understand how television fantasies serve people, a good way to begin is by examining dream fantasies.

Dreams are produced by almost everyone almost every night. From midnight to dawn a vague but lavish parade of happenings and personages, episodes and colors, marches through minds. The ornate and nonsensical images rise almost to the level of conscious memory before subsiding again. Much of this material, when it can be recollected, seems based on recent events in the dreamer's life. Investigators report that a dreamer will typically feel a quarter of his dreams are unpleasant, and another quarter are neutral or mixed, but almost a half are experienced as truly enjoyable.

The wellspring of dreams was identified by Sigmund Freud back at the turn of the century as the deep-lying mental level of the unconscious. His observations of patients in his psychoanalytic practice led him to conceive of the unconscious as a kind of holding tank where those impulses and animosities which should not be openly displayed could be contained. Some of this repressed energy was thought to come from primal drives which to a greater or lesser extent churn in all minds, as humans genetically retain much of their animal past. Also held back in the unconscious were retaliatory feelings which couldn't be outrightly expressed. An individual might feel an imposition from someone in authority—a parent or boss—but would recognize that striking back would only lead to greater hardship; the anger would be restrained in the unconscious.

Although Freud's concept of the unconscious is derived from his treatment of troubled patients, it appears applicable to everyone. Male or female, adult or child—no one can let loose with everything that pops up in the mind. For all humans, the process of growing up in society is the process of learning how to rein in these impulses. It's indisputable that the human brain has some capacity for constraining inappropriate urges.

But when a person sleeps, and the usual sanctions can be relaxed, then whatever has been sentenced to the unconscious can be vented in dreams, to the relief of the individual. The sensation of pleasure which was often reported to Freud along with accounts of dreams convinced him that psychic pressure was being discharged. Basing his ideas on the research of early 20th-century physiologists, Freud held that pleasure was a sensation which occurred whenever tension was reduced. In dreaming, the strain of repression was giving way to the pleasure of release.

Fantasy is the means by which the unconscious is relieved in this nightly activity. At the start, a dream fantasy may closely resemble the dreamer's real world: the settings are places he has been to, the characters have the names and appearances of people he knows, the events resemble events he has personally experienced. This high degree of familiarity means that the dreamer is partially bound up in the fantasy; his presence and feelings are woven into the dream's texture. But as the fantasy proceeds, it may become clear that some of the dreamer's personality has been left behind. What he does not bring to the dream are many of the obligations and inhibitions that govern his everyday behavior. In the dream fantasy he is allowed to do things and feel things that he usually cannot—to take chances, try on other personae, rub up against bizarre carryings-on. He can be playful without feeling foolish, just as he can be lecherous without experiencing guilt, or vengeful without suffering recriminations. All's fair in a dream fantasy.

In the "what-if" world of fantasies the dreamer acts out the contingencies of the mental pressures behind the dream—sometimes over and over, until the psychic tension is finally dissipated. If the pressure has been directed in the fantasy to a forceful conclusion, catharsis can be the result. Whew, the dreamer may feel at the end.

Central to Freud's view of the mind were these twin concepts: that there are limits to how much repression the unconscious can retain before personality becomes distorted, and that letting off some of this psychic pressure through dream fantasies is normal and therapeutic. Freud knew that for many people in his times, dreams were held in low repute. In a lecture delivered when he was visiting the United States he explained, "Our low opinion of them is based on the strange character of those dreams that are confused and meaningless, and on the obvious absurdity and nonsensicalness of other dreams. Our dismissal of them is related to the uninhibited shamelessness and immortality of the tendencies openly exhibited in some dreams." Freud went on to assure his audience that dreams were "compatible with complete health in waking life." No matter how grotesque they might be, there was nothing more human and healthy than dream fantasies.

Modern-day psychologists have not had cause to revise Freud's idea that, when too much is forced into the unconscious, the pressures can bulge up and influence lives for the worse. Bruno Bettleheim, in his award-winning book on the utility of fairy tales, *The Uses of Enchantment,* explains why the unconscious must be regulated: "In child or adult, the unconscious is a powerful determinant of behavior. When the unconscious is repressed and its content denied entrance into awareness, then eventually the person's conscious mind will be partially overwhelmed by derivative of these unconscious elements, or else he is forced to keep such rigid, compulsive control over them that his personality may become severely crippled." Nor have modern-day psychologists seen any reason to question Freud's belief in the good that fantasy can do. Dr. Seymour Feshbach, head of the Psychology Department at the University of California at Los Angeles and a nationally known researcher into fantasy and aggression, told me during the course of an interview, "Fantasy is one way people get rid of tension. The better a person is at fantasizing, the less need he'll have for other avenues."

Deprecated in Freud's day, dream fantasies have now come to be regarded positively, or at worst neutrally; presently it's television fantasies which often receive scorn and condescension. But no matter what value is associated with them, it's objectively the case that television fantasies serve humans in the very same way that dream fantasies do. This correspondence is implied by the fact that, when individuals are wired to an electroencephalograph, the brain wave patterns during both television-viewing and dreaming have been found to be similar. The dreamer and the viewer experience a decrease in fast beta waves and an increase in the slower alpha waves. Neurologically, television seems to have the same effect as dreams.

One person who recognizes that television is an extension and amplification of the dream world is Dr. Edmund Carpenter—an anthropologist by training, a communications philosopher by inclination, and a sometime collaborator with Marshall McLuhan. In his speeches and writings Carpenter has stressed the parallels between viewing and dreaming. "Television extends the dream world," he says incisively. "Its content is generally the stuff of dreams and its format is pure dream." Carpenter explains that television does not record the real world but rather "a world within." He states, "TV, far from expanding consciousness, repudiates it in favor of the dream."

Carpenter would have no difficulty understanding a recent report on a Cree Indian village located in the northern reaches of Canada. Like many primitive people, the Cree feel intimate with a spirit world which lies just out of view. So that things go as they ought to and catastrophe is fended off, the Cree try to keep in touch with the otherworldly forces they feel swirling about them. Dreams are the way these humans and the spirits communicate with each other. Through a ceremony known as "shaking tent" ("koosabachigan" in the Cree language) the reception of dream-messages used to take place. A shaman would crawl into a special tent, close the flap behind him, and go into a trance. When strange voices began to issue from inside and the tent skins started to quiver, the villagers clustered around the edge knew that the spirits were being received.

The Cree are among the last people on the North American continent to be reached by television. By means of satellites they began getting signals in the mid-1970s. Interestingly, the new technology was also labeled with the word "koosabachigan." The ancient shaking tent ceremony quickly fell into disuse as the Cree turned to their television sets for messages from afar. An anthropologist living among them observed a bit dejectedly, "There is evidence that the Cree now turn to TV to study the programs for omens, for dangers, and for directives concerning proper behavior pathways into the future, once provided by dreams." One Cree respondent told him straightaway that the original television had been dreams.

Television is a kind of dreaming for the Cree, and for us, and for everyone else. Just as in a dream fantasy, a television fantasy will feature behavior that, if we stop to think about it, is not possible in the real world. No household offers as many laughs as Archie Bunker's did, no squad room overflows with decency like Barney Miller's, no one actually acts like Mork or Kojak or the Hulk or Wonder Woman. And just like dream fantasies, television fantasies too usher out pressure and tension from the unconscious and help people feel relieved. According to survey researchers, the sensation of relief and relaxation is what Americans say they get from the medium—the same sensation that Freud's dreamers reported to him.

The difference between dreaming and television-viewing is that in the first case fantasies are supplied from within the brain and in the second they come from outside. To judge from the size of the television audience and the amount of time it spends with the medium, broadcast fantasies may be felt by Americans to be an improvement over the fantasies they dream up themselves. This is suggested statistically by ratings given the two kinds of fantasy: Americans find only about half their dreams to be truly enjoyable but fully 80 percent of the television shows they select. There are several reasons why television fantasies might be preferred to dream fantasies.

The dreams of humans are frequently not dramatic, while the supplemental fantasies provided by television aspire to be—and are with surprising frequency. Unlike a dream, a television show is a structured whole whose careful unfolding can be intriguing to a large number of people. If all goes as it should, each segment of the show will fit in with the others (with a logic that is emotional, not rational), and the members of the audience will be carried along the plot line. The dramatic pattern of television fantasies is an age-old one: characters are introduced; something complicating and thwarting occurs; the story builds toward a climax through a series of intertwined events; finally everything is set to rights again. This familiar form elicits the commitment of the audience in the expectation that whatever feelings are stirred up at the outset, they will be replaced with pleasurable ones at the end.

The characters in television fantasies are different from the people that wander around in dreams. The raw material for dreams is everyday life, so the personages that populate them can be thoroughly ordinary. Characters in a television show, on the other hand, are likely to be quite uncommon, of an elevated sort that members of the audience would be inspired to rise toward—either to identify with or respond to.

There are other aspects to television fantasies than tried-and-true structures and fetching characters that may make them more effective and enjoyable than dream fantasies. In dreams people often reexperience the debilitating emotions that plague them in the real world— loneliness, self-consciousness, lack of resolve, and the like. In television fantasies there is less chance of meeting these ample, minor feelings, and more chance of being exposed to the strong emotions which are indemnifying yet all too scarce in real life—frequently, those of dominance and affection, whose starkest forms are the violence and sex that Media Snobs deplore.

Similarly, in dreams individuals have to work with the physical and social settings they confront in their real lives, with material which is so close to home it's possible the fantasizer won't be able to let go to an extent that's truly pleasurable. The conventional binds that a person may wish to escape can still be around him in his own dream fantasies. But a television fantasy can transport an audience away from all that, to locales and coteries that viewers may find exotic and liberating. The setting can't be so far outside the audience's frame of reference that it becomes difficult to enter into it imaginatively. However, short of that there is a broad territory people can be invited into, as they are also invited to leave behind some of their usual inhibitions.

If the setting is attractive for the audience, and the characters strike fire, and the depicted emotions are ones yearned for, and the plot is well timbered, then the television fantasy may serve as no individual dream fantasy can. It might be logical to think that a person's own dream would be better able to relieve his particular mental pressures than a mass media fantasy, but this does not seem to be true. A Hollywood-created fantasy, because of its dramatic qualities, can cut more incisively into a mind and release more harbored feelings. Whatever has been repressed will be more completely eliminated.

It is reasonable that television fantasies be more effective than dream fantasies. They are the product of an enormous amount of talent and application, having been created by professionals who have years of experience at converting the public's vague longings into vivid images and stories. In New York and Los Angeles these experts labor intensely to fabricate the most appealing possible fantasies. The pressures of the television industry, and of their own personal ambitions and pride, decree that their show be as competitive as possible. They may not realize they are competing with dreams, but they are, and to some extent they are winning.

Charlie's Angels was a prime example of this sort of created and televised fantasy. The show featured three young women working as operatives for a private detective who was strangely anonymous. He issued them their assignments and orders by telephone and through an intermediary. Against criminals the girls were more than capable of giving better than they got, so everything was always set to rights at the end. But along the way there was enough tussling and mild violence to keep viewers alert. And enough bra-less chests under scanty clothing to make it a girl-watchers' field day.

The chest that initially received the most attention belonged to a new actress, Farrah Fawcett. A woman with an engaging smile and trim figure, she struck a chord with the audience in a way that few television performers do. *Time* magazine referred to her as a "spectacularly maned frosted blonde," and so singled out her hairdo, as did many among her multitude of fans. There was something exalting and inspiring about her head of hair, and soon girls across the nation were imitating the side curls that framed Farrah's face. A poster of her torso and magnificent waves sold more than any poster ever.

With Farrah as the star, *Charlie's Angels* rocketed to the top of the Nielsen ratings as soon as the show debuted in September 1976. It captured 60 percent of the audience in its Wednesday night time slot (when 30 percent is all a show needs to stay alive). A storm of letters from its new audience was loosened, reaching 18,000 weekly. A poll of high school students taken in 1977 at the end of *Charlie's Angels'* first season showed how closely Americans had drawn toward the fantasy. Asked which female they most wanted to become like, the girls put Farrah at the head of the list. The boys selected Lee Majors of *Six Million Dollar Man* presumably not so much because they admired the man or the show, but because of his marriage to Farrah in real life. She was a star in every teenager's fantasy.

Trying to account for the remarkable popularity of the show, former producer Rick Husky said, "What we're talking about is a B exploitation show. Not even a B. We understood we needed to exploit the sexuality of the three girls, and that's an obvious reason for its success." He may have reason to disparage the dramatic quality of *Charlie's Angels,* but when the standard of fantasy is applied the show ranked among the best. The program appealed to men and women alike, and to their differing needs for fantasy material. Males could identify with the omnipotent Charlie, who in effect had a harem at his command. On the few occasions where Charlie and his life-style were glimpsed, viewers got the impression of an active, randy man. Charlie was an invisible but omnipresent person throughout each episode, having a relationship with the story line not unlike that of a dreamer to a dream. For the males in the audience, who at times in their lives are far more powerless than they wish to be, to be godlike and watch over the doings of a triad of beautiful girls is an extraordinarily enticing fantasy. Females could identify with the girls, who were first of all attractive, then energetic and assertive, but still under the aegis of a dominant male figure. It is not a fantasy of feminists, but it is one that suited the majority of American women in the late 1970s, having the right mix of protection and venturesomeness.

After what amounts to a long production run in the quicksilver world of broadcasting—some 109 episodes in total—*Charlie's Angels* was finally cancelled in 1981. Its sex and violence had been piquant enough to attract the reproving attention of the latest reform group, the fundamentalist Coalition for Better Television, and when the show began to falter in the ratings these Snobs, not unlike wolves trailing the feeble and the stray, had gone after it. At the final press conference Aaron Spelling, the coproducer, observed, "The critics would come and discuss the serious aspects of the show. 'How could you have the girls fire a gun and blah-blah-blah?' I have always said this, 'If you really believe that there are three gorgeous girls who work for a man they've never seen, who's a voice over the telephone, then you've missed the point of our tongue-in-cheek attitude.' It was a fantasy to start with."

Televised fantasies like *Charlie's Angels* are indeed an amplification of Americans' dream worlds. Once the truth of this solution to the mystery of television-viewing is recognized, then the six clues we began with make sense.

1. Television-viewing is a personal activity, not a social one, in the same way that dreaming is a personal activity. There is nothing more confidential than the lower reaches of the brain where one's direst feelings seethe; this is the place where dreams and programs alike attach themselves. The functions of television are more private than those of the bath.

2. Television-viewing is enjoyable because psychic pressure is being eased. Simply put, the store of repression and tension is being depleted and the viewer experiences relief. It's a pleasure to shed some of the unavoidable accumulation of frustration and resentment in a manner devoid of guilt or reason for retribution.

3. Television-viewing is needed because the capacity to store psychic tension is a limited one. A person can hold so much and no more. Unless the pressure is going to be vented in antisocial ways, fantasy material will be in demand.

4. The casualness of viewing is due to the fact that it is the unconscious regions of the mind, not the conscious, which are the target of television images. Viewing does not have to be closely done, and probably should not be, if it is to have the desired effects. It can be as haphazard as the unconscious processes themselves. By viewing absently and piecemeal, the content may have the best chance of seeping down into minds and dissolving the debris.

5. Television-viewing is logically an evening activity because then it can help reduce the day's buildup of tension. It is the antithesis of what goes on during working hours. Because it succeeds at unwinding people, television has displaced the traditional activities which did this.

6. Fraternizing, sleep, other innocent diversions—over the last three decades they have surrendered a bit to the intrusion of television. People don't chat with their neighbors and family quite as much as they used to. A midwestern lady interviewed for a *New York Times* article on the disappearance of front porches volunteered, "I remember we had a porch with those old creaking wicker swings and chairs and we used to sit out there at night like this and chat with passersby and see who was

going where. But today you've got television. I tell you I'm terrible for it. I turn it on in the evening after a day's work with people and I can be relaxed and entertained. It's easier to watch TV." Along with almost a quarter-billion other Americans, she had found television to be the great relaxer. Perhaps her bedtime, like most others, had been pushed back to make room for still more viewing. Television does rival sleep and dreams at ridding brains of tension.

What these items add up to is this: the true purpose of television is to remove mental debris from the minds of viewers. The medium can be seen as the sanitation man of the psyche. Its fantasies are permitted to enter the minds of humans and rumble around, picking up all the flotsam and jetsam they can. Once the trash has been collected, then away the fantasies go, out of mind. They work so well, and cart the refuse off so distantly, that the viewer can barely recall they have been in his mind—if a researcher asks him what he viewed the night before, he is hard put to remember. But the viewer does sense that he has been left somewhat refreshed and repaired.

Like all fantasy material, television does not so much put things in brains as it does take things out. Television does not add to our mental stores; it subtracts from them.

Fantasy and Reality

To grasp the significance of television's fantasy world, it has to be contrasted to the real world where we spend most of our lives. The real world is the place where schooling goes on, work is done, social relationships are made and maintained; it's the place where effort counts. The majority of us are successful in the real world the better part of the time; if this were not true, society would cave in. But for all our successes, and even for our merely enduring, there are psychological costs. Mental strains are the price paid for real-world exertions. What fantasy does is to relieve the distress that results from our day-in, day-out tussle with reality.

Present-day reality may be especially formidable, increasing the need for the salve of fantasies. There are two aspects to existence today which make it more arduous and treacherous than that of our forebears: their lives were governed by convictions and certainties which are elusive nowadays, particularly with regard to social roles; and the pressures of the workplace have continued to mount.

People in previous decades and centuries had very firm senses of what was to be expected from men and from women, of what parents were to do and how children were to behave. Such assuredness seems almost luxurious from our perspective. The development of industrial life has brought about a dissolution of traditional social roles and a confusion of new, half-formed ones. This is interestingly revealed in a survey of advertising pitches made over the last century, compiled by Robert Atwan and his associates. Not so very long ago advertisers could make contact with consumers simply by invoking their needs to be homemakers, if female, or breadwinners, if male, says Atwan. Nowadays advertising appeals are much more diffused, reaching toward what are depicted as more independently minded women and more carefree men. Modern-day individuals are people who have to feel their way along without the bracing support of strong social definitions.

The most fundamental area of life to be negotiated is that of employment. Increasingly Americans work for large corporations which are locked into ceaseless struggles against other large corporations. All these companies, seeing their payrolls as a major outlay, are determined to get the most out of their employees in order to remain competitive. The employees, in turn, are striving to succeed within the structure of the organization. Their ambitions drive them to accomplish as much as they can. The upshot is that there is an enormous amount of pressure on each American worker, pressure that comes from both outside and within. Most Americans go home tired out at the end of the workday.

Aaron Spelling, who in addition to *Charlie's Angels* also produced *Mod Squad, Fantasy Island,* and many other well-known television series, is a man whose wealth has elevated him from the usual pressures of the workplace, but when he explained why the fantasies he produced were so popular with the American public, he said, "You can take me out of being very poor and make me very, very rich, but I still know that when someone works all day with his hands as my father did, he comes home and uses the TV set as a paintbrush to paint over the horrors of the day, to forget what real life is. I am giving people a happy pill in the nicest sense of the word." It's not just blue-collar workers who need Spelling's "happy pill"—it's white-collar workers, too, right up to the pinnacle of management. Viewing records disclose that nearly all Americans like a dab of fantasy after a day of stiff reality.

In order for televised fantasies to be therapeutic, they must in several senses be the opposite of the real world they are remedying. It's in the nature of an antidote that this be so. But Media Snobs, failing to appreciate the true function of the medium, are often galled by the lack of similarity between the television world and the real world. The scientific literature on the social effects of television is laden with numerous studies in which the characters in broadcast dramas are tallied and their distribution is compared to that of the real world. Invariably the findings are that the proportion of male characters on prime-time television greatly exceeds the proportion of males in the real world, and that the occupations represented on the medium are so highly skewed toward crime-fighters that one would think it's the sole career for adult Americans. The authors of these studies are left to grumble about what they think is a distorted picture of American life conveyed by these programs. This line of research would make much more sense if television put things into brains instead of taking them out; then there would be some reason to complain about the poor fit between televised fantasy and the real world. But since television entertainment does little to teach about the real world, and much to compensate for it, then it's reasonable that the characters would oblige fantasy needs and not instructional ones.

The television characters who are the best vehicles for the audience's longing for escape from workday constraints tend to be adult males, since they remain the most likely type to assert themselves upon their surroundings. In the national attitude survey conducted for the 1972 Report to the Surgeon General, favorite characters mentioned by respondents were virtually all adults and mostly male, even though one-fifth of the sample were adolescents and more than half were female. Edith Efron of *TV Guide* realizes that the audience's desires, and thus the networks' profits, require male characters: "Network TV's economic existence, and an enormous number of its technical calculations, are totally based on the certain, *proved* knowledge that the overwhelming majority of the U.S. public and its kids is fixated on the simple, continuous vision of good, just men. To a striking degree, network TV's profits flourish in the loam of hero worship." Burley assertive crime-fighters are ideal vessels for viewers' fantasy demands because they are both laudable and aggressive, able to act as angrily as viewers may momentarily wish to, but without fear of recrimination since they are lashing out legitimately against evil.

Just as these males are not representative of real-world males (few of whom are pistol-packing, quick-witted, depression-free, and always right), so are television females quite different from their everyday counterparts. In the real world women come in all sizes, shapes, ages, and dispositions, but on television they tend to be of one kind. The actresses who find work in the medium can play women who are young, proportioned in a certain way, attractive, and with a perky but not assertive personality. Feminists find much to be irritated about in these stereotypes, but clearly the fantasy needs of Americans—both male and female viewers—are being fulfilled.

Television characters and their doings may be more exciting than people in their everyday lives, but in another respect televised fantasy is tamer than the real world. While reality can be uneven and full of surprises, our nightly entertainment is highly regular and patterned. Most shows come in the format of a series—a structure developed in the radio days and taken over completely by the newer medium. Although the broadcasting of film and mini-series has provided some relief to this scheme, the series mode is network television's framework, and promises to continue being so.

Within each series, most episodes are highly formulaic, with the same characters going through largely predictable patterns leading to expected resolutions. To some extent this simplicity results from the limitations imposed on production by schedules and resources. With almost five million hours of programming broadcast annually in the United States, there is a barely sufficient pool of writing, acting, and producing talent available. Accused of mediocrity, one station manager was prompted to say, "Hell, there isn't even enough mediocrity to go around." There are other factors which also keep each program formulaic. The episodes cannot take up as much time as they might need to satisfy interior dramatic demands, but must end exactly at the half-hour or hour mark. Subtleties can be lost for no other reason than the technology's small screen size and poor resolution. Subtleties will also be wasted on an audience known to be inattentive.

But even if all these were somehow corrected, television programs might still be formulaic. What viewers have desired so far is predictable material, material which would not tax them too much as they looked for the fantasies which helped ease out tensions. They have wanted familiar characters in familiar formats because these are the fantasies which served them before. Popular culture analyst William Kuhns says, "Audiences won't usually choose the unexpected over the expected, and the series orientation has provided the most reliable basis for satisfying expectations so far." Kuhns has formulated this into his Guaranteed Expectation Principle: people watch what they know they can expect.

Kuhn's Principle has its antecedents in Sigmund Freud's "repetition compulsion." Freud knew that people sometimes have the same dream repeatedly, night after night, and that children frequently want to hear a familiar story read to them, although they know the ending full well. For both adults and children an unvarying fantasy occurs in response to a particular psychological pressure which it can help control (although not permanently eliminate). Freud saw this at work in the instance of his 18-month-old grandson who had invented a fantasy game that he played whenever his mother went out. The game seemed to help the child realize that disappearance did not mean abandonment. It is not a great leap from this kind of repetition to the repetitiousness of television shows. Viewers want to see familiar fantasies: a new series should have a kinship to an old series, and once established it should keep the same characters doing the same sort of things.

Non-serial broadcast fantasies, such as specials and made-for-television movies, often achieve this requisite familiarity in another way. The raw material of these one-shot productions tends to come from themes and events that have already intruded into the public consciousness, thus giving the audience some foreknowledge of the subsequent fantasy. During the '70s more and more of the content derived from topical issues like homosexuality, abortion, minorities, teenage runaways, skyjackings, nuclear protest, child abuse. Some observers of the television industry say Hollywood organizations turned to this material because other, more psychological themes had been used up in the enormous volume of programming, but another explanation is that producers recognized the American audience had preliminary awareness of such topics, and that this would help in the coupling of the viewers and the fantasy.

This tendency is carried to the extreme in docu-dramas, where real world events have been hammered into dramatic forms. The viewer of *Missiles of October* or *Helter Skelter* or *Kent State* vicariously participated in historical situations which had been refashioned into fantasies. It is not the smack of reality which the viewer wants from such programs; it is fantasies which are familiar enough that it is emotionally easy to enter into them.

Being the reciprocal of real life, the fantasies on television help compensate for the world's burdens. Bob Shanks of ABC, one of the shrewdest observers of the television phenomenon, comments, "Television is used mostly as a stroking distraction from the truth of an indifferent and silent universe and the harsh realities just out of sight and sound range of the box. Television is a massage, a 'there, there,' a need, an addiction, a psychic fortress—a friend." The real world can unbalance us and bow us down, but for most people prime-time fantasies help to restore psychic equilibriums.

All kinds of Americans, in all states of mind, turn to the medium for the balm it provides. The most troubled are perhaps the most aided. For the segment of the population that has been crushed by the real world, and has had to be removed from it, television is clearly a boon. Anyone who has visited an institution where humans are confined knows that television exerts a calming, beneficent influence. Walking down the corridors when the sets are on, sometimes no noise is heard other than what is being broadcast. The administrators of hospitals, prisons, and asylums realize that their charges can be highly volatile or depressed, and that television is an efficient, nonchemical means for easing their torments. Inmates say that taking television privileges away is among the crueler punishments.

A series of scientific studies has documented that, for people who are functioning in the real world but are distressed, television helps bring peace of mind. In a 1959 study published in *Public Opinion Quarterly* by L.I. Pearlin, examination of 736 viewers revealed that those with higher levels of stress and anxiety reported the greatest satisfaction with television entertainment. A similar study was conducted in 1967 by W.R. Hazard at the University of Texas and printed in *Journalism Quarterly*. Hazard gave 430 adult viewers a standard psychological test of anxiety, and also asked them, "If you could plan a perfect evening's television, what shows would you be most likely to see?" His conclusion was that the most anxious people were the most likely to desire fantasy programs. A 1977 study of sixty-four households by two University of Minnesota psychologists found that those individuals who were the most tense had the highest likelihood of turning to television for redress. Psychic pressure and television fantasy go together like problem and solution.

This is true for the most troubled and for the most anxious, and it is also true for the majority of sane, normal Americans. All who are subjected to the real world come away with a certain amount of stress in their minds. It could hardly be otherwise, since all adults must wedge their way among the obstacles that comprise reality, and each child must suffer through the restraints of being reared. In every human mind there is some repressed energy, whether it be anger, frustration, strain, hurt, impulse, or whatever. It is only natural that people would want as much of this pressure removed as possible. They want the fantasies into which this seething energy, the poisonous dregs of our time in the real world, can be poured.

Television delivers the fantasies which millions upon millions of well-balanced, capable Americans use to relieve the daily accumulation of mental pressures. The restful and satisfying feelings they receive from their dosage of video fantasies are the first thing they tell survey researchers when asked their overall judgment of the medium. Herbert Gans, a Columbia University sociologist and media scholar, conducted a poll of television viewers in 1968, asking among other things, "Do you ever turn on the TV to help you get over feeling blue or in a bad mood?" The majority of the respondents said that they did. Over 80 percent said they had felt "especially good or cheerful because of a TV program they watched." They said this, talking to a strange poll-taker in a doorway about their private needs. If they had been chatting with a confidant, the figure might have been even higher for the bulk of Americans who sense that television is a good answer to tension.

Two More Kinds of Content

A general theory incorporating much about what viewers get from the media has been formulated by Gerhardt Wiebe, dean of the School of Public Communication at Boston University. What started him thinking along these lines, he said, was the disparity between the enormous audience for television on the one hand, and on the other the content which critics agreed was trivial. More than met the eye was going on here.

Wiebe theorized that for an individual the process of playing a role in society could be sorted into three phases: 1) *directive,* 2) *maintenance,* and 3) *restorative.* He explained, "The first includes learning, refinement, improvement in the direction of prescribed behavior. The second includes the relatively stable, acceptable, everyday behavior of one's achieved level of socialization. The third includes the retaliatory, assuaging, indemnifying counterstrokes." Wiebe then went on to examine the mass media in terms of *directive, maintenance,* and *restorative* messages.

The greatest number of media messages, he said, are *restorative*: "The *restorative* aspect of socialization is served copiously, though not of course exclusively, by the kinds of media content that seem so deplorable to those with discriminating taste. *Restorative* media messages feature crime, violence, disrespect for authority, sudden and unearned wealth, sexual indiscretion, freedom from social restraints. The themes of these most popular media messages seem to make up a composite reciprocal of the values stressed in adult socialization." Wiebe concluded by saying, "The *restorative* mechanism hypothesized here has as perhaps its chief merit the characteristic of releasing hostility in small amounts." Pent-up tensions are alleviated by the fantasies which Media Snobs seem so set against.

In Wiebe's account of the importance of television entertainment in socialization, he touches on the services provided children as well. Children are the ones whose behavior is most molded and rearranged in the interest of meshing with society: thus they are greatly in need of *restorative* material if they are to maintain their psychological equilibriums. This suggests that television fantasies can have even greater utility for children than for adults.

Regarding *directive* messages, a small but definite appetite for news and information does exist. About one quarter of the adult American population will watch television news on a typical weekday and accept facts about the state of the real world. The series of Roper polls regarding audience attitudes indicates that by 1970 television had become the most popular source of news, felt to be most credible and most complete. Yet it is also the case that this reality material is thought to be relatively unimportant and incidental. Only 10 percent of a national sample mentioned information-seeking as a reason for watching television.

In between the *restorative* content of television fantasies and the *directive* material of television news lies the broad middle ground of *maintenance* messages. The types of programs which belong here can be a mixture of all three categories. Game shows, for example, can be *directive* in that they convey new information, however trivial, but game shows also permit people at home to identify with a winner and triumph, and so experience *restorative* feelings. Yet fundamentally game shows are a facsimile of the real world, where gain and loss occur in roughly equal proportions. Because game shows rehearse the viewer's sense of how the world works, they are essentially *maintenance* programming.

Similarly, talk shows belong in this middle range between fantasy and reality programming. *Maintenance* content, with a dash of the *restorative* and the *directive* for good measure, is the gist of the morning-hours shows like *Today* and *Good Morning, America* and of the talk shows like *The Phil Donahue Show* and *The Tonight Show with Johnny*

Carson. All these programs transmit a certain amount of digestable information—some of it new like the day's weather or the title of a just-released movie but much of it of an unspectacular, confirming nature, the sort of cultural ballast everyone needs to stay on even keel. In addition, these shows put forth attractive personalities which many viewers could imagine themselves interacting with; in the sense that they beckon us from our situations in life, talk show hosts and guests are offering Americans escapist, *restorative* fare. The female viewer can link arms with her loveable hero, Phil Donahue, and go rocketing off into unexplored terrain, playing Lois Lane as Donahue charges into new ideas, practices, norms. But talk shows are chiefly an enhancement of the viewer's world as it exists, neither calling for new responses nor producing the cathartic feelings that can accompany fantasies.

The most prominent variety of *maintenance* programming, however, is soap opera. Like the real-world information which television sometimes brings, the function of soap operas is to put things into viewers' brains and to add to what is already there. But like the entertainment shows, these particular programs are constructed of fantasies. Soap operas flesh out viewers' psychological lives usually by supplying fantasy groups which viewers can imagine themselves belonging to. While most Americans most of the time suffer from an excess of interpersonal contact, at moments there is not enough, and loneliness threatens. This can be especially true for those who stay home all day, and for the increasing proportion of the population which lives alone. Soap operas, then, are ways of extending and peopling a sparse social universe.

Overall, the content of the medium amounts to this: about 10 percent of what the American audience wants from television is information about the real world; another 15 percent consists of the *maintenance* programming, primarily soap operas, which augment private lives; and the entire remaining 75 percent are the fantasies which deplete the reservoirs of tension and help restore psychic well-being. These last are the shows the audience aches for. In survey after survey viewers report that entertainment is virtually the only reason they turn on the set. The extent of this need was amply revealed by an event that occurred in 1966. Gemini 8 with astronauts Armstrong and Scott aboard was in grave danger, and all three networks interrupted their programming. NBC received 3,000 complaining telephone calls in protest. The bulletins that interrupted ABC's *Batman* were the object of 1,000 complaints. As the lives of the real-life astronauts hung in the balance, CBS was besieged by callers denouncing the cancellation of a science fiction potboiler called *Lost in Space.*

A view on reality is not what we generally want from television; we want the antidote for reality. As Edmund Carpenter says, "TV is actually a *blind* medium. We may think of it as visual, recording a world 'out there.' But it records a world within. Sight surrenders to insight, and dream replaces outer reality."

A Question

What Americans want from television is the dream-like and formulaic fantasies that will vent mental pressures. But how does it happen that we get what we want? We certainly don't get what we want in the real world; why should it be that we get what we want from our favorite pastime?

4

How It Happens that We Get What We Want

Feedback

Let's consider this event, even though at first it looks inconsequential. One wintertime afternoon in southern Illinois, a high school junior skipped basketball practice because of a head cold and flagging spirits. Under leaden skies he walked the short distance home. No one else was there—his sister had band practice, and both parents were still at work. His father managed an auto parts store, his mother was a bank teller. He turned up the thermostat and, after taking off his coat and laying his books on the dining room table, went into the kitchen. There he made himself a sliced cheese sandwich and poured out a can of his mother's diet soda. Plate and glass in hand, he walked into the chilly living room.

Before sitting down he turned on the television set. Then he closely watched what came on.

That's all there is to it. Except that he belonged to one of over a thousand "Nielsen families," as people at the A.C. Nielsen Company like to call them. These are the households that comprise a national sample of television viewers, and whose channel selections are under Nielsen's magnifying glass. When the boy switched on the set, an electric impulse saying that Channel 7 was being received traveled to a gadget, an Audimeter, in the front hall closet. There the information would remain until later that day when a computer over a thousand miles away, in an office building near the bright, warm coast of the Gulf of Mexico, would send out a signal over special telephone lines. The signal called in the day's viewing record from the Audimeter, and from over a thousand other ones scattered around the country. All his family's station choices, from the moment he turned on the set just before 3:30 in the afternoon until the end of Johnny Carson's monologue when his father snapped the set off many hours later, rushed from the Audimeter and across the nation to the Nielsen computer in Dunedin, Florida.

The Media Research Division of A.C. Nielsen collects data from the Nielsen families daily, tallies it and arranges it for the benefit of their clients, and then periodically sends it out to automatic printers located in television networks, program suppliers, advertising agencies, and advertisers here and there around the country. These reports are the Nielsen ratings of the popularity of television shows. They tell the television industry which shows are relished by the American public and which shows are barely tolerated.

The flow of information from the high school student's home in Illinois, and from the thousand other households, into the Nielsen offices, and then on to the networks and agencies, is the primary form of feedback in television. From it media executives learn how well their programming is being accepted. They discover if *Laverne and Shirley* has more viewers than *Little House on the Prairie* and how David Brinkley compares to Dan Rather. This boomeranging curve of communication—network to audience and back to network again— influences upcoming programming decisions as television personnel try to keep after the fickle, elusive audience.

Every communications system, big or small, must have feedback if it is to avoid falling apart. This is true for a simple system like two people conversing. As a person talks, he is also looking at the listener to see how well his words are going over, although he may be unaware of his own watchfulness. By being in the presence of the eyes and expressions of the other person, the talker perceives interest or inattention, and if he is like most people, modulates accordingly. Feedback, often in the form of nonverbal signals like winces and nods, helps him target his conversation. If he ignores the feedback he may shortly lose his listener. The same principles applies in the mass media. Those who send messages, if they are to continue sending messages, must have a heightened awareness of how well the messages are being received. This is very important in a competitive situation, where the audience can turn to another broadcaster who better understands their likes and dislikes. But even noncompetitive media, as in the Soviet Union, still operate with feedback. The attentiveness of the Russian people, or the lack of it, is enough to let those in charge of the communications apparatus know when to change their ways. Communist leaders will do what they can to avoid a restless citizenry.

Soviet media was a special interest of Harvard University social scientist Raymond A. Bauer. Through interviews with Russian workers Bauer learned that—contrary to what many Americans thought—the Russians made highly selective use of the media and were not dupes of the government's propaganda. By choosing what they would or would not pay attention to, the citizens were being active participants in their nation's communications systems. The more Bauer thought about this, the more he became convinced that the wants of an audience are an important ingredient of the mass media anywhere. In an article titled, "The Obstinate Audience," published in *The American Psychologist,* Bauer explained that the willfulness of the audience had a decisive effect upon the content of mass communication. It puzzled him that so many academics saw the audience as listless recipients of a one-way stream of communication. The media, he wrote on the basis of his own and other studies, entail "a transactional process in which two parties each expect to give and take from the deal approximately equitable values." A transaction, an exchange, a two-way flow of communication—that was the better way to picture how the media work.

Networks and audience are linked by an endless circuit of messages. The networks send messages to the audience in the form of television programs, and the audience sends messages back via the Nielsen ratings. Depending on the audience's retorts, the networks will adjust their offerings for a better aim. So it goes, on and on, round and round. If information did not circulate in both directions, the system would come tumbling down. The networks would have to create shows in the absence of a good sense about what the audience wanted, and soon would discover that no one was watching their shots in the dark. The audience would turn to other diversions that served it better. But since feedback does occur in the form of Nielsen's numbers, the television industry can learn what the public wants, and can send out programs which attract a sea of viewers every night.

The fact that television programming is so finely attuned to the desires of the audience is in good measure due to the attentiveness of the networks as they vie to hear what the public is saying. ABC, CBS, and NBC are locked in battle, each trying to beat the other in the act of converting feedback on present programs into next year's shows. The struggle goes on without letup, season after season. The people who work in the television business are victims of unrelenting strain. The pressure is always on, no job is secure for long, the bars neighboring network headquarters in New York City fill quickly after 5:00 P.M. No broadcaster can get enough of the audience to be satisfied. It's not even profits that tell who is winning and who is falling behind in this attention-grabbing business; it's popularity. Although CBS had record sales and earnings in 1977, President Robert J. Wussler was shifted out of office when ABC managed to dislodge the traditional leader from first place ratings. Audiences, not earnings, are the key to long-term success.

Not heeding the ratings, and not aiming for general popularity, has caused the Public Broadcasting System in the United States to stumble through the years. PBS is acknowledged by all but watched by few; it averages 2 percent of the viewing audience. George Comstock, in his authoritative *Television and Human Behavior,* commented, "Proportionately, the audience for public television is so small that a description of the viewing behavior of the American people would not be seriously distorted were no mention of it made." The upshot of having a tiny viewership is that financial support has turned thin, while the ratings-guided networks have become fat and sassy.

The jostling among the networks to obey the whispered commands of the audience conforms to idealized theories of competition but is nonetheless exceptional. Many industries in the United States are governed by a tacit agreement among the member companies to avoid headlong competition and to divvy up the market like gentlemen. Feedback from the consumers is often treated casually since none of the producers is too anxious to change its product and alter the status quo. Everyone is content with a less-than-frantic marketplace. Not so in the television industry, where the networks bump and bang each other in their zeal to appease the public. As long as the scuffling continues, the audience remains the beneficiary. Our favors are curried as if we were potentates. Platter after platter is held up before us. Even as we may be saying, "No, I don't want that," our hands go out and we begin indulging.

The Nielsen Ratings

The Nielsen ratings were around even before television arrived on the scene. The idea of tabulating listeners' station choices evolved as radio evolved. It was a partial response to advertisers who demanded to know what they were getting for their dollars. If they were going to conduct their advertising campaigns rationally, they needed some sort of comparative tally, such as they had managed to force on newspaper and magazine publishers in the preceding decades. Media scholar Hugh McCarney explains, "While print publishers and theater managers could measure consumption of their product through the number of copies or seats sold, radio was free and the message disappeared as it was sent. Before radio became an advertising medium the number of listeners was a matter of curiosity, but after 1922 when WEAF (in New York City) began to experiment with selling time for commercial messages, advertisers needed to know how many people were being reached." Various ways of calculating the audience were tried out—interviews, coupons, postcards, telephone calls—but each new method led to more doubts and reconsiderations. How do you best count a traceless audience? What really is being counted? If you have a count, what use can you legitimately make of it? Schemes were proposed by people who knew little about measurement and social statistics, but who sensed a business opportunity. It was a wonderful chance for fraud, and radio audience measurement in the '20s, and '30s had its share.

As pressure began to build for a purely objective head-count, some of it flowed in the direction of a Chicago-based marketing research firm, A.C. Nielsen. Founded in 1923, Nielsen had among its clients firms like Campbell Soup and Johnson Wax. The audience size for their new radio advertising was a matter of concern to such clients, and they encouraged Nielsen to devise a way to measure it. Arthur Nielsen began to search around. An engineer by training, he was more inclined to mechanical devices which would keep track of listening than he was to dealing with the say-so of respondents. When two MIT professors came up with an instrument which could be attached to a radio and would record when it was switched on or off as well as any changes in tuning, Nielsen quickly acquired the rights in 1936. For the next three years the company worked on the development of the service, and by 1940 Nielsen was issuing radio ratings.

The service was costly, and as a result the Nielsen ratings were less in demand than those of its competitor, C.E. Hooper. Hooper used the more economical telephone-coincidental method, where a flurry of random telephone interviews produced estimates of what the public was listening to at the moment. After the Second World War Nielsen stayed with radio, while Hooper raced ahead to start up television ratings. But the Hooper company turned listless during the slack period of the 1948–1952 freeze. When Nielsen offered to buy out the Hooperatings in 1949, Hooper was losing not only money but also faith that national television would ever catch on. In March of 1950, for a half million dollars, Hooper sold his national rating services, telling friends that he had unloaded a worthless property. Nielsen immediately became the company with the best national figures, a prominence it has never lost. Profits came in due course. The competition between Nielsen and Hooper was a case of the tortoise outdistancing the hare.

All told, it took 17 years of development and 15 million dollars invested before the A.C. Nielsen Company could break even on its Audimeter service. Companies without this sort of determination and capital fell out of contention long before. Nielsen succeeded in part because it was making enough money in its other operations that it could survive the drain of development costs. The firm was the world's leader at checking supermarket and drugstore shelves to see how well their clients' products were selling—a service of great importance to manufacturers who have to learn as quickly as possible about any snags in their long distribution channels and about how their sales compare to the sales of others. Income produced by the shelf-checking operation helped tide the company over as it became obvious that developing the ratings business was going to take time, since there were few forerunners and Nielsen would have to carefully feel its way.

It is an irony that the Nielsen ratings, which have come to count so much in the operation of modern media, count so little today in the operations of the A.C. Nielsen Company. Measuring television audiences amounts to only 11 percent of this large firm's business. Yet in the world of television, those ratings numbers are the be-all and end-all. "In television," writes ABC vice-president Bob Shanks, "everything depends on the question, How did it do in the ratings?" *New York Times* television writer Les Brown echoes this when he concedes, "In television only one notice matters, that from the ultimate critic, the A.C. Nielsen Company." The Nielsen ratings are the conduit of the public's choices, and when they arrive at the network offices, it is with considerable authority.

Although everyone calls them "ratings," what A.C. Nielsen terms "ratings" may not be the most telling information the company has to offer its subscribers. A "rating," to Nielsen, is the percentage of all households with television sets installed that are tuned to a particular show. But obviously some television households won't have the set on during that period. If you were a television producer and you wanted to know, from among all the households which actually had the set going, the percentage that was tuned to your show, then you would want to know what Nielsen calls the "share." "Shares," not "ratings," are the real ratings because they are more of a competitive measure.

Shares and ratings are computed by Nielsen from the daily summation of the one thousand Audimeters (the name is a holdover from the radio days—Videmeters might be better for these times). Considering how crucial the Nielsen families are, it's surprising that Nielsen pays them so little for the privilege of invading their homes. Each family gets $25.00 when the Audimeter is installed, $2.00 a month thereafter, and half of all the set's repair costs. Nielsen says that paying more would only distort viewing behavior.

The magical thing is that the one thousand Nielsen families are a representative sample of the millions of American viewers. When we stop to think about it, sampling such as the Nielsen organization undertakes is easy enough to fathom. We all sample every day of our lives. A person puts his nose outside the door in the morning and takes a sample of the weather. From it he thinks he can generalize about the weather in his locale and even say what it's going to be for the next little while. Someone goes to a newly opened supermarket and from a sample of ten purchased items, out of the hundreds available, judges whether the supermarket is more expensive than others or less. If people didn't sample they would drive themselves crazy because they would have to experience everything before they could reach a conclusion. They would have to drink a whole quart of soured milk rather than accept the information of a sample sip.

People in the sway of Media Snobbery sometimes state they don't understand sampling, or how the Nielsen families can stand for all American families. There are several reasons why they might say this. The findings of the Nielsen survey often contradict the desires of them and their friends—they feel allegiance to one kind of programming, and the ratings tell of the popularity of others. In their minds the Nielsen figures exist only to obscure the fact that the networks are going to send out unwanted shows no matter what. Some Snobs object to the sampling because they are uneasy about the incursion of a marketing firm into the privacy of the American home—they sense something sinister, not scientific, is going on. As much as representatives from A.C. Nielsen try to explain the nature of sampling and the reason why the thousand chosen families are a fairly good index of most Americans' viewing preferences, a large number of vocal skeptics remain. An exasperated Nielsen executive once said that the next time those people went for a blood sample, they should tell the doctor to take it all.

Taking it all, or doing a complete census, may not be an improvement over taking a sample. It is possible for a census to be less accurate than a sample, if only because it is very difficult to contact everyone who is supposed to be reached. The 1970 United States census, heir to two hundred years of thinking about how to count Americans accurately, was off by an admitted 2.5 percent—a worse error margin than Nielsen concedes. The 1980 census also had problems. A census is not a guarantee of a perfect poll, while a sample might be more exact and in any case is certainly less expensive.

Since so much rides on it, the Nielsen sample is concocted with great care. The process begins with official population figures. The Census Bureau counts people but it does not list them since this would be an infringement of their rights. Instead it lists census tracts, or neighborhoods. Nielsen statisticians pick random census tracts from master lists, being sure to duplicate the geographic and social distribution of the American population. The percentage of small town tracts in the sample should equal the percentage of small town tracts in the United States today, like the one where the Nielsen-tallied high school student lived in southern Illinois. Next, a Nielsen representative drives out to the selected tract and at random picks a house. Through the luck of the draw a new

Nielsen family is set up. (But the one family alone represents nothing. It is a misunderstanding to say, as Art Buchwald did one time in his newspaper column, that when one Nielsen family goes to Grandma's house for dinner, 60,000 go. Only the total, aggregated sample has meaning and represents anything.) Those who agree to join up are told to keep it a secret, and apparently they do, because over the years very few have betrayed Nielsen's trust and publicly identified themselves as a Nielsen family. Since the selection of a Nielsen family and the installation of an Audimeter takes time and money, the sample is not changed much. Nielsen says there is a 20 percent turnover annually.

The Nielsen sample consists of a minute fraction of all American households, but because the sample is so carefully crafted, that fraction is enough for obtaining a satisfactory picture of the whole audience's tastes in television. If there were fewer families in the sample, the possible error range would be unacceptably high. If there were more, the gains in accuracy would be offset by the costs of data collection. Precision does not increase directly with increases in sample size, which means that many more households would have to be counted for even the slightest improvement in accuracy of measurement. Hugh McCarney, in his study of rating systems, summarizes the situation: "The television household sample size that seems to give the degree of accuracy that ratings subscribers are willing to pay for in a national sample is about 1,000."

The accuracy of the Nielsen ratings became a matter of federal interest in the wake of the quiz show scandals, and the early 1960s saw a series of Congressional and industry-backed investigations into sampling and rating. It was widely believed that the scramble for good ratings was the cause of the quiz deceptions; the investigators went after not the scramblers but the raters. The findings were that the rating services, Nielsen chief among them, were generally accurate, although not perfectly so. Nielsen's sample at the time was based on a 1947 research design which had become outdated by later advances in statistical and probability techniques. Nielsen was told to revise its sample, and with some fanfare it did.

A convincing demonstration on the power of sampling was made to the Congressional investigating committee by the Committee on Nationwide Television Audience Measurement, an offshoot of the National Association of Broadcasters. From an earlier study they had inherited a collection of some 56,000 television viewing diaries. Ratings from the 56,000 households had been computed. Then samples of 500 and less were plucked randomly from the 56,000. It was shown that even when the sample was as small as 50, the resulting figures were almost identical to the ratings that had been computed from all 56,000. The Congressional committee was impressed. They had been led to see sampling as an economical and accurate method for gathering data about large numbers of people. Everyone seemed to emerge from the hearings with more faith in the ratings, including Nielsen, which ran an even prouder ship after this point.

Detractors of the ratings system are still quick to point out possible flaws, however. For one thing, there is no attempt to measure non-household viewing—in bars, college dorms, hospitals, institutions. Even the household count is less than ideal. For years Nielsen has been criticized for not fully tabulating the viewing of minorities, and even now, by Nielsen's own admission, the number of black families in the sample is below the proportion of black families in the United States census. Additional sample bias stems from the fact that just 70 percent of the households which the Nielsen representatives approach allow the Audimeter to be installed. What are the characteristics of the three families in ten who refuse? Do they have traits which mean that segments of the American population will be underrepresented in the Nielsen sample? It is believed that the "noncooperators," as the rating companies call them, tend to be politically conservative, less well educated, less well salaried than their neighbors who accept the Nielsen invitation. So, contrary to what Media Snobs may believe, the Nielsen sample is probably skewed toward the better educated and more affluent segments of the public. In any case, it is definitely inclined away from the poor and institutionalized.

Nielsen hears few complaints from its subscribers about this bias. It fits in with the subscribers' own slant on marketing and the media. They are after the best customers, the ones with coins jingling in their pockets, and those are the ones that Nielsen finds easily. Ratings then are not purely democratic, only approximately. The furthest margins of society have been pared away.

Another lingering problem highlighted by critics of the system is the question of exactly what the figures are measuring. By saying "household tuning," Nielsen has obviously chosen a shorthand for audience size, which is what media executives want to know. But how good is the match between "households" and the actual number of viewers? And what is the relationship between "tuning" and viewing?

It used to be easier in the old days when there was rarely more than one set per household. And indeed, when households in the traditional sense were more in evidence. But households have become smaller in recent years, while television sets have proliferated. The shrinking number of viewers for each set suggests that Nielsen overestimates the audience, since for ease of accounting it considers each of its metered sets to be a "household." The three-set family counts as three "households." Another development: as television sets have become more common, viewing has grown less intense, to the point of touching zero at times. How much of the Nielsen ratings are accounted for by sets whose electrons are madly racing through their tubes, with sound and image blaring out, all this to an empty room? There is no way to tell. As Hugh McCarney says, "It is dangerous to assume that tuning and viewing are the same thing."

One mismeasurement in the Nielsen figures we can be sure about is that they report inflated hours of viewing. Nielsen families cannot help but be self-conscious about their viewing; an outcome of this is that their sets are on more than those of most people. Leo Bogart, in *The Age of Television,* estimates they watch 20 percent more than average. But like the skew in the composition of the sample, subscribers to the Nielsen ratings don't worry too much about these puffed-up figures. Since they are in the business of selling audiences to advertisers, the larger Nielsen says the audience is, the more money they stand to make.

Nielsen is not measuring how much viewers are receiving, but how much television sets are receiving, which is a different thing. The chief liability of the Nielsen ratings would seem to be that somehow, curiously, humans are missing from the tally. No attempt is made to assess the real headcount, or the real number of hours people spend in the company of the set, or the real extent of their receptivity. Is the whole family avidly watching *60 Minutes,* or is the canasta club talking above it, or is the 8-year-old alone and playing with her dolls? All Nielsen knows is that the set is on and tuned to CBS.

And yet, paradoxically, this very liability—the missing humans—is also the strength of the Nielsen measurements. Other measuring systems which depend on human interviewers and human respondents have not worked out so well, even though they continue to find some application in local ratings because they are comparatively cheaper to use. Humans are odd, value-ridden creatures. They value privacy above openness, high status above low, ease above effort, and many things above objective fact. They will fill out viewing diaries before the viewing week has begun, tell the telephone interviewer the set is on when it is known to be broken, claim to have closely watched specials that were never telecast, swear their children only view a few minutes weekly, refuse to acknowledge their spouse's football games or soap operas. Measuring humans means dealing with values and character quirks, with momentary moods and whims. By minimizing the human element, Nielsen is able to arrive at objective, standardized figures which its clients can make ready use of.

Let's assume—and there is good reason to—that Nielsen's sample mimics the whole population of viewers, and that "tuning" is the best available way of estimating viewing. Snobs could point to still other possible problems with the Audimeter data. To keep its prices down, the Nielsen organization has settled for a very large margin of possible measurement error. They can only claim their ratings are exact two out of three times. Most social science researchers are unhappy if their accuracy falls below the level of nine out of ten times. Nielsen maintains its figures are actually more precise than the two-out-of-three range because it makes repeated measurements during a viewing season, which permits it to draw close to (but still not meet) the operating standard for most research. Thus the ratings are not only estimates, they are somewhat rough estimates. In addition, Nielsen can make outright mistakes —and has done so. In 1975 Nielsen first announced that television viewing was off, but afterward higher figures were issued. (Maybe they were right the first time, because measurements in subsequent years confirmed a leveling off in viewing times.) The January 1978 ratings were incorrect by several points for several shows, red-faced Nielsen officials confessed when corrections were released several months later. They took pains to say the error was not caused by the computer, but by a computer operator. Since one rating point can be worth one million dollars in total advertising revenues, personnel connected with the misappraised shows were irate.

But when all is said and done, the television industry maintains its faith in the work of the A.C. Nielsen Company. The figures may not be perfect, but they are satisfactory for comparing the audience of this show to that show, this network to that network, this season to last season. Whatever mistakes Nielsen may make, they are inadvertent, everyone believes. The Nielsen organization is thought to be uncompromisable from within or without. There was an attempt to influence the ratings several years back, in 1966, but it was caught by Nielsen security. About 6 percent of the Nielsen families received in the mail a form designed to stimulate their interest in viewing a particular network on a particular night. It did this by asking them about certain commercials to be broadcast on February 18. Tracking these letters back to their source, it was learned that they had been sent out by one Rex Sparger of Washington, D.C. Sparger turned out to be a protege of House majority leader Carl Albert. Somehow during the Congressional investigations of the rating services Sparger had managed to obtain at least a portion of the highly secret Nielsen list. Trying to explain, Sparger lamely said that he wanted to expose the rating system and gather material for a book. But later it was discovered that Sparger had accepted $4,000 from the husband of show business personality Carol Channing. *A Night With Carol Channing* had been broadcast on February 18. This was the last widely reported instance of tampering with the national ratings.

On balance, then, the Nielsen ratings are honest if crude indicators of the tastes of the American audience in television programming. Through them the public's likes and dislikes are telegraphed to the networks. If the ratings did not exist, feedback would be much more paltry and erratic, and our chances of getting what we want would sharply decline. Anyone who has turned on a television set in Hungary or South Africa or Singapore knows how unappealing the content can be when audience measurement counts for little. But owing to the open, coursing feedback of the Nielsen ratings, the voice of the audience is heard distinctly in the United States. No other mode of feedback can do as well—if 50,000 letters were written about one show, that would represent less than two-tenths of one rating point. It is through the ratings that the circle is closed in the two-way communication of audience and broadcasters.

It may seem precarious that the immense communication system of television depends on such a slender thread as the Nielsen ratings. All that booming programming goes in one direction, and only the single filament of figures goes back in the other. Unfortunately, it isn't practical for there to be more than one set of rating numbers. Michael Wheeler, who closely and skeptically examined the Nielsen ratings in *Lies, Damn Lies, and Statistics,* says why: "Each network cannot simply produce its own estimates of its viewing audience, for such data obviously would be suspect. Thus it is in the networks' interest to have a single, ostensibly neutral scorekeeper to report on what programs the American people are watching." So it all hangs on the Nielsen ratings. In spite of whatever shortcomings they might have, they have proven themselves up to the task. It may even be because of their shortcomings that they are so sturdy—the ultra refinements of sampling theory and interviewing and data compilation that have complicated and even distorted other polls are simply omitted. The robustness of the Nielsen service, the result of cautious early development of the Audimeter and of circumscribed ambitions at measurement, is what allows the ratings to perform so well.

The American public has a deep-lying sense of the importance of the Nielsen ratings. When a favorite show goes off the air, fans will write Nielsen headquarters in Northbrook, Illinois, and blame the company for the cancellation. They are blaming the messenger for the bad news he carried. Invariably, Nielsen sends back a well-worded if form-letter explanation to the effect that it didn't take the show off the air—the network did. If it wanted to tell the whole truth, it should say that neither it nor the network is guilty—the viewers are. The smoking gun is in our hands.

The Sponsors

The networks can't be content to base their programming solely on the Nielsen reports of what the audience used to like, weeks and months before. The ratings may tell which shows to drop, but they don't tell what shows it would be best to start up. Networks must strain to look ahead and anticipate what the mercurial public is going to long for in the season that is to come. This crystal-ball work is difficult, preposterously so, but it is a responsibility networks have to accept if they are going to stay on top of the situation. Some media executives are better at this than others, and in just a bit we'll look at the man who for several years was the best of all. But first we have an obligation to consider what ought to be the primary force in television, and which so far we have shunted to the side. This is the party that picks up the tab for the shows—the sponsors. Since they are the ones that pay, isn't it likely that they influence the content of television? Do they subvert the choices of the audience?

There is good reason to think that advertisers do affect programming. Their interest and the interests of the viewers are not identical. Advertisers are so bent on getting people to buy more that it is hard to believe their venality does not slop over into the programming, and that they do not interfere with the mechanisms of audience feedback in order to get programming that benefits their own agenda. Television would be bringing us not what we want, but what advertisers want us to see.

Advertisers certainly pump more than enough money into television to give them the right to shape the content. Over $12 billion flow from advertisers into the television industry annually. To rent time for a 30-second commercial can cost well above $100,000 in prime time. A 30-second spot during a football game can extract over $200,000 from an advertiser. But it is not just the cost of broadcast time that sponsors pay for. Through their advertising agencies, they must also pay exorbitant prices for the production of the commercials that go into the slots. A 30-second commercial can cost up to $300,000. On the average the figure is five times what it costs to produce the same amount of prime-time entertainment, so meticulously are the commercials made. "Commercials are more carefully prepared, more elaborately produced, and more frequently seen than any one program on television," states Jeff Greenfield in *Television: The First Fifty Years.* Although the average life of a commercial is only about six months, during that time it will be seen by more people than have watched every theatrical production in America for the last 100 years. For an audience of those dimensions advertisers want it done right, no matter what the cost.

The power of advertisers to control programming in the medium which they sustain used to be clear-cut. In the early days of television sponsors paid for, and often produced, an entire show. Program titles— the *Philco Television Playhouse, Kraft Television Theatre, Colgate Comedy Hour*—announced who was in charge of the content. Depending on which way the sponsor leaned, the show was sure to follow. In the cold political winds of the '50s, the tilt of the sponsor was the tilt of the program. Amm-i-dent Toothpaste caved in to the anti-Communist crusades and compelled the networks to purge shows of left-leaning content and personnel. In the other direction, Alcoa sponsored Edward R. Morrow's *See It Now* and kept the show on the air during the exposes of Senator Joseph McCarthy.

With the sponsor being so conspicuous in programming, no one was really surprised at some of the silliness that went on. If the Ford Motor Company wanted to have the Chrysler Building erased from the New York City skyline as the camera panned across it, technicians in the film developing studio were ready to oblige. If the gas industry, sponsor of a *Playhouse 90* production about the Nurenberg trials, wanted all references to the gassing of Jews expunged, a production assistant was happy to bleep them out as the show was telecast.

In his book *The Sponsor,* Erik Barnouw testified in the traditions of Media Snobbery to the evil hand of advertisers in small matters like these, and in large matters as well. From the outset, he said, advertisers contorted television to their own ends. "That programming generally plays a business-supportive role—at least avoiding anything at odds with commercials—is scarcely noticed by most viewers." Entertainment, he felt, was the mode of television that was pushed by advertisers, because once viewers had been diverted and critical faculties were at rest, the sponsor could slip his message across unimpeded.

The television business has changed radically since the beginning, however, and the power which advertisers exhibited then they have now largely relinquished. Although it seems far-fetched that corporations would uneventfully, even happily, surrender such great power, that is exactly what has transpired. The reason that it has happened so smoothly is that it suited the interests of everyone concerned, the sponsors most of all. Advertisers rapidly lost interest in tying up their resources in the flighty and fast-changing television business. But most decisively, their advertising strategies changed, and changed so completely that program control became far more of a burden than an asset. Originally they wanted to broadcast their messages over and over again to each potential customer, so they would purchase an hour on radio or early television which every week at the same time would attract roughly the same audience. People would buy the advertised product in part because of the "gratitude factor"—a phrase much heard in media circles. Unfortunately for the sponsor, most of the audience turned out to be ingrates.

As television began to expand after the 1948–52 freeze, it became clear that its best feature was not its ability to badger the same people with the same message week after week, but its potential for reaching vast numbers of new, ready customers. Advertisers began to see that the most productive use of the medium was to place commercials throughout the broadcasting week, and not to tie themselves to one show in one slot. According to media observer Martin Mayer, it was Procter and Gamble that led the way to scatter buying. A media buyer for the detergent company said, "If you're in twenty shows, you can't be hurt. If you spend it on five shows and two go bad, you're in trouble." But more than that, a much higher percentage of the market can be reached with twenty shows than with five. Reach, rather than frequency, became the battle cry of television advertisers. Interest in being totally responsible for any one show and its content plummeted.

At the same time that advertisers were retreating from program control, networks were feeling more ambitious, and were thinking of a larger role for themselves than merely that of broadcast technicians. They began to take command of the programming, and to implement the desires of the advertisers. Reach was what the advertisers wanted, and reach was what the networks would provide. Weak shows, even those with strong sponsors, would have to go if they did not preserve gargantuan audiences so that there would be "audience flow" from one hour to the next. Out went *Voice of Firestone,* over the protests of the tire company. But most sponsors were glad to accept the new way of doing things, for the networks were much better than they were at attracting the multitudes.

Actually, it was not each and every American that advertisers wanted networks to round up for them. It was almost but not quite each and every American; some were to be ignored because the advertiser knew they were not good customers. This selectivity—not extensive, but definite—is to this day the major way that advertisers exert an influence on television content. Advertisers will only pay for commercial time on shows that are likely to attract potential consumers. *TV Guide* writer Richard K. Doan says, "TV programming is governed by the law of survival of the fittest. The fittest being the shows best liked by certain people. Certain people being those the advertiser wants to reach. The people he wants most to reach being, as a rule, the great well-washed middle class, 18-to-49-year-old group. Young adults, they're called in the trade. Kids and older folks the sponsor can ordinarily do without— unless he's selling cereal or Geritol." Thus *Star Trek* was cancelled because its fans, while fervent, were too young, and advertisers had shyed away. It had the wrong "demographics," media personnel would say. Similarly, shows with preponderantly elder viewers do not have attractive demographics, so *The Lawrence Welk Show* was let go. By focusing on the mid-range American, advertisers cause the television industry to produce programs with the population mainstream in mind, to the disadvantage of subsidiary groups of less-than-good customers.

This amounts to a passive, almost inadvertent influence of advertisers upon programming. In the beginning advertisers wanted to determine the totality of the programs, but in the end they recognized that their goals were best served if they remained aloof. "Advertisers are to regularly scheduled television what the Queen of England is to the British Commonwealth," cracks ABC's Bob Shanks. The networks are intensely aware of their patrons, and respectful of them, but proceed in the conviction that only they know how to come up with the programming that will garner mass audiences. Advertisers and advertising agencies are perfectly willing to stand aside and wait to see which producers and which networks will have the shows that best meet the intentions and budgets of their campaigns. Only infrequently do they intrude. "While advertisers can have an influence on programming," observes media scholar Hugh McCarney, "it is primarily informal and takes the form of suggestions."

Once again, it all boils down to the ratings. As one advertising agency executive said, "The client wants to know how many people he's reaching with his TV money, and ratings are all we have to go on." The better the ratings, the more people are exposed to a commercial, and the more favorable the association is between the advertised product and a winner of a show. High ratings are what advertisers want, seek out, and pay for. From the networks' point of view, high ratings translate unambiguously into high profits. In the 1974–75 season, to pick one year as an example, CBS beat NBC by only one rating point, but that was worth $17 million in revenues, about 85 percent of which was pure profit. "Having an advertiser-financed broadcasting system works toward directing the broadcasters' attention to objective audience analysis, and making the commercial system more responsive to popular taste than the noncommercial systems," comments McCarney. In American television, it is the taste of the public which network and advertiser accede to. The audience determines what television carries.

Fred Silverman

Converting audience feedback into programming is the *tour de force* of the television industry. For years the best in this business was Fred Silverman, who is as close to a folk hero as the networks have yet produced. When Silverman hopped from one network to another, newspapers announced it and magazines displayed him on the cover, financial analysts quivered and network stock prices fluctuated by several points. What did he do to deserve such attention? In the eyes of Media Snob Richard Reeves, Silverman "puts into practice H.L. Mencken's dictum that no one ever went broke underestimating the intelligence of the American public." Seen another way, by television producer Norman Lear, "He is a showman, perhaps our P.T. Barnum. Freddie is decisive, courageous, and smart—all are rare in television."

Fred Silverman was television's paramount programmer. Like a strategist in a war, a network programmer tries to position his troops along the battlefield of the weekly schedule so that he has the best chance of triumphing over the enemy. Or plural enemies, in the case of television, where each network has two others to combat. In the battle of the media, Silverman was Patton and Eisenhower and MacArthur rolled into one. The reason for his victories, lightly commented television writer Marvin Kitman, was "Freddie's solid grasp of the fundamentals: what works once (*Six Million Dollar Man*) will work twice (*Six Million Dollar Man* and *Bionic Woman*) and what works for a half hour (*Happy Days*) will work for an hour (*Happy Days* and *Laverne and Shirley*). Add to that Freddie's ability to play the schedule like an accordian: some shows he stretched out (12 hours of *Rich Man, Poor Man* ran for 10 weeks), others he squeezed together (*Roots,* 12 hours long, ran on eight consecutive nights)."

There was more to it than duplicating winners and toying with the schedule. In the development of new programs, Silverman was always in the thick of it, pestering the producers, directors, and writers until he got what he thought the audience would find attractive. He might consider 1,500 series proposals until he got down to the handful that would be worked on; he read virtually every script and every rewrite of every script for the shows, pilots, even potential pilots. These nascent programs were the raw recruits of his annual campaign, and he wanted them to be as fit as possible. When he was with CBS he was credited with everything from *Green Acres* to *All in the Family,* from *Hee Haw* to *The Autobiography of Miss Jane Pittman.* During his brief stint at ABC he brought in *Charlie's Angels, Three's Company, Laverne and Shirley, Happy Days,* and *Soap.* At NBC came *Sheriff Lobo, Hill Street Blues.*

So great was Silverman's fame that for a time many believed he could not only build mighty schedules, he could also singlehandedly rescue networks. When he went to ABC in 1975 it was trailing far behind, and when he left two years later it had shot ahead to first place. Twenty-one stations had defected to ABC, 15 from CBS and six from NBC. In 1977, the four highest rated shows, and seven of the top ten, belonged to ABC. Its prime-time average Nielsen rating of 21.5 outstripped the 18.7 figure for CBS and NBC's 18.0. The lead was worth $100 million in advertising revenues to ABC.

With hindsight we can detect the roots of his success. The son of a Roman Catholic mother and Jewish father, Silverman was born in New York City in 1937, and grew up as television grew up. There was something unique about his family that foreshadowed his career—his father was among the first television repairmen. The ambivalence of many Americans about the new technology was little heard in the Silverman household, we can imagine. After high school Silverman left New York City and went upstate to Syracuse University, where he received a Bachelor's degree in Radio and Television. Then on to Ohio State for a Master's degree in Communications in 1959.

Fred Silverman's Masters degree thesis is interesting reading, suggesting much about the 22-year-old and something of what was to come. Fifteen years before he was to save ABC for real, he wrote over 400 pages on "An Analysis of ABC Television Network Programming." It is an authoritative study of ABC programming as the network stumbled through the '50s. In the interests of survival the weakest network had stuck with the most vicious shows, something the young, hard-nosed analyst found acceptable. "ABC's entrenchment in the action form was quite explainable—it would have been sheer foolishness to attempt a totality of program service comparable to the CBS or NBC schedules; not with blue-chip advertisers Liggett and Myers, P. Lorillard, Procter and Gamble, American Home Products, and several others willing and eager to join the ABC ranks with purchases of the action/adventure programming." The critics be damned: "Television critics and pasteboard programmers have all the answers on paper though they fail to realize that network television is basically a business, with profit and loss columns, stockholders meetings, and annual reports."

This was a young man with a highly developed sense of the realities of network television, its programs and audience. What is of special significance is that, although presumably all media topics were open to him for his thesis, Silverman chose to bring his mind to bear on the problems of the network that was running last. There was no affinity for losers here; the interest was in analyzing a redeemable situation, as a prelude to struggle and triumph. Many years later Silverman said, "I think there is a philosophy that is good no matter what you are doing. That is to always act as if you're in last place. You just shouldn't take success for granted." It was the motto of a determined man.

Following graduate school Silverman worked briefly at a Chicago station before he was hired by CBS in New York. At the age of 25 he became the head of daytime programming, and seven years later he was given full control of the schedule. CBS continued to be the front-runner during his reign, and ABC continued to lag behind. In the middle of the 1974 season it became clear to ABC that something would have to be done or it would fall out of the running altogether. The organization turned to a new president, Fred Pierce. Resolved to hire the best programmer, Pierce sought out Silverman, whom he had first met when Silverman was writing his Ohio State dissertation.

How much credit could Silverman take for the near-miraculous turnaround at ABC? Not all of it. Much was due to Fred Pierce and other executives who, as ABC sank lower and lower, caused it to rebound and ascend all the way to the top. A network headquarters is a large and complicated organization, and it takes more than a chief programmer to make it run smoothly. Still, the programmer is in a position to be the single most important person on the staff, and Silverman made himself that at ABC. The network was poised to rise; the addition of Silverman meant it could. When he was subsequently lured away by rival NBC in 1978, it was a difficult parting. "He swore he would not leave us for a competitor," wailed one ABC man. But for a million dollars a year and the presidency of the network, Silverman moved on.

Once at NBC, though, things turned sour for Silverman. The network's ratings and profits were on the skids, and when he rebuilt the schedule in mid-season, it only accelerated the decline. The following year the new shows he had overseen didn't take with the audience: one on the subject of marriage, aimed at educated, reflective viewers, *United States,* failed; *Skag,* a series with dramatic pretensions which starred Karl Malden, failed quickly; *Supertrain* failed spectacularly. Silverman forecast that NBC would be tops in the ratings by Christmas 1980, and then had to recant. In the spring of 1981 higher management in the parent RCA Corporation released him from his duties.

As to why Silverman had succeeded at CBS and ABC the keys to his success were several. His Master's thesis discloses his powers of analysis; he has an extremely acute mind. One thing Silverman added to this was the will to work longer and harder than anyone else. Bob Wood, once president of CBS, said, "I never worked so hard in my life as when Freddie was working for me at CBS. He knows what goes into every pot, just like a chef." Everyone who had been in league with Silverman commented on his application—this is an industry where hard work and long hours are standard. Jeff Greenfield quotes a colleague of Silverman's: "He had plans, charts, cards. He'd broken everything down. He was totally compulsive. He'd go over a script or story line again and again and again. Someone would say, 'We just did that, Fred,' and he'd say, 'Let's do it again.' "

Silverman differed from many other television executives in that, by all accounts, he truly enjoyed viewing. "He had the most contagious enthusiasm of any executive at that level I have ever seen," reminisced an ABC staff member. His laughter or tears at previews were legendary. A CBS higher-up recalls being invited into Silverman's office to look at a soap opera episode. "It was a routine hospital bed scene, with the man standing beside the bed of the woman he loves. But I looked over at Freddie, and tears were rolling down his cheeks."

Silverman's real secret was his ability to be not just the sender of television programs, but an unabashed receiver as well. A rival remarked, "It's not that Freddie understands the audience—he *is* the audience." Having the tastes of the viewers he was trying to reach, Silverman was able to short-cut the feedback process to some degree. His competitors' shows can tend to be hit-or-miss, while his shows, guided by his gut feelings for the audience, were frequently on target. "The Man With the Golden Gut" was the title of a *Time* magazine cover story on Silverman, which observed, "What makes him the best programmer in television today is the fact that he is the best viewer working in television today." He was the audience's premier representative in the network offices.

The shows that Silverman searched for in our behalf were those with the sort of main characters to whom viewers could feel a special link. He said, "I think the thing that makes a successful show an enduring show is well-delineated, attractive, appealing characters." This was his viewpoint from the beginning; in his thesis he predicted that *The Untouchables* was going to be a front-running show, as indeed it became, because "it has a leading man with whom the audience can identify," Robert Stack. "Without this audience identification, the most elaborately produced television would be a spectacular failure." He adjusted *Kojak* to fit the personality of Telly Savalas, rather than the other way around. "He doesn't believe in premises as much as people. He'll suggest this actor, that actress," said a producer who had worked with him. A talent of Silverman's was to identify the personalities that would somehow strike fire with the viewers. At CBS he picked out William Conrad for *Cannon,* saw Bea Arthur on *All in the Family* and proposed *Maude,* viewed Sonny and Cher only once before signing them, shoved *The Mary Tyler Moore Show* into the prime slot on Saturday night after glancing at the first rough cuts of the actress's performance.

A typical Silverman show when he was at CBS or ABC would weave these characters into a group with strong personal and family bonds, at home or work. Of all that he did at CBS, Silverman is proudest of *The Waltons. All in the Family* and *Mary Tyler Moore* survived season after season because of the strengths and rewards of the human contacts among the characters. It is the sort of warm social milieu which Americans are neglecting in their go-it-alone lives, but still thirsting for.

What then happened when he got to NBC? For one thing, the audience was changing—it was maturer and more hard-pressed than the carefree, novelty-seeking viewers of a few years before. Constant schedule changes and fledgling shows were not what they wanted. Also, the company was in a less salvageable position than ABC had been, with lower morale and worse relations between network headquarters and the local stations. The loss of the 1980 Olympics was a stiff blow to the organization. But most importantly, Silverman himself was a different person. Passing on into his forties, he had become a devoted family man; this may have taken the edge off his intensity. His feel for programming was further blunted by his new, elevated role, since the president of a broadcasting company has many other responsibilities competing for his attention. With one thing and another, Silverman lost the knack of catching the public pulse. He and the audience parted company, and his time as America's programmer was at an end.

What's next for Fred Silverman? Jay Sharbutt, an Associated Press television writer and humorist, thought he had an answer. After NBC, Freddie would go on to head up the Public Broadcasting System. He would start off with three blockbusters—*Downstairs, Downstairs, Easthampton Beach Bum,* and a drama about ancient Rome called *I, Laverne.* Sharbutt conjures an old-timer at PBS who says, "It could mean the end of the dull documentary as we know it."

Viewers' Choices

Among the speakers at the week-long convention of the National Association of Television Program Executives in 1978 was Ted Turner, America's Cup yachtsman and the owner of an Atlanta television station. Outspoken as always, Turner said, "We've been doing a lot of talking about quality this week. Let's not kid ourselves, most of the stuff we put on the air is garbage. High ratings and big dollars are what we're interested in. Quality—that's how much money we make off it." He got the biggest hand of all.

If we strip away the confessional and exorbitant tone that makes

Turner good news copy, and expose the basis of his remarks, perhaps we too can applaud. The television business is not keyed to the tastes of a few who have their own notions about "quality," he was revealing, but to the greater number of viewers, whose choices are reflected in the ratings. Turner may choose to call those choices "garbage"—that's his right. But they are, first and foremost, choices that have been made by the public-at-large. The choices have to be honored if "big dollars" are to be earned. No matter what their personal feelings might be, broadcasters are compelled by the economics of the situation to be attentive and responsive to the wishes of the audience. If they are not, revenues will shift to their competitors.

The driving force in the television industry is the desire to do better at giving the public what it wants. As Martin Mayer says, "Once a network's income became a function of the ratings of its shows, the tendency to seek the highest possible audience for each minute became a compulsion, irresistible, ultimately seen as 'natural.'" The highest possible audience with the right demographics, that is. We shouldn't forget that the choices of some viewers are not counted—if you are a college sophomore, or a buck private, or a convict, no one cares what you are watching. And also, some choices that are counted are later thrown out, if they come from households with low purchasing power. It may be that those who are purposely slighted by the television industry are not really offended in the long run, for statistically they are the heaviest viewers. In any case, with television, as with few other things in this world, the great majority of us get what we ask for.

One reason Media Snobs maintain this is not true is that for viewers without cable hookups the menu is so small, it seems improbable people can really get what they want. There are only three networks, after all, not a dozen. Half tongue-in-cheek, network executive Paul Klein says that with so few offerings, viewers do not select what they might ideally want but settle for the least noxious. This is Klein's "Least Objectionable Program" theory, or LOP. There would be more to the idea of LOP if it were not for the fact that behind the current threesome is a history of thirty viewing seasons. Tens of unappreciated program types have been jettisoned, and only what the audience likes best has been retained. Viewed historically, the field of choice proves to be far larger than three. (LOP is also undercut by the fact that, of all media participation, television-watching is the most purposive and selective. People listen to the radio or read newspapers much more rotely than they view television.)

It is not accurate to say though, as a blanket statement, that audience feedback is the only influence upon programming. Other forces are sometimes at work too. One is the Federal Communications Commission which, because it is the chief government overseer of the media, would seem to wield a great deal of power. But the same government that can tax and conscript and execute is very reluctant to challenge the Constitutionally guaranteed rights for an unhampered flow of communication. The Commission can interfere with the routines of relicensing stations, but rarely does. Almost all the decisions which it has made pertain to purely technical matters. When it strays into the area of programming, it can leave things worse than they were before. In 1970 the Commission adopted the "prime-time access rule" whose practical effect was to open up the 7:30 to 8:00 P.M. slot to non-network program suppliers. The upshot was that game shows, the cheapest and easiest of formats, increased from one-tenth to two-thirds of the programming for that period. Such results make the FCC shier still about trying to influence television content. Pressure groups may appeal to it, but the response is always similar to the one given by Chairman Richard Wiley in 1977: "I tell people at PTAs and other group meetings that your government does not have the authority to do what you want it to do. And if it did, do you realize that a government which has the power to deal with excessive violence and sex on television could also influence news content and other aspects of programming?"

Local stations, poised between the networks and the audience, can be another influence on programming, and not only through the shows they might produce themselves or rent from syndicators. If a program is sent out from network headquarters which they think is not right for the local audience, they may refuse to carry it, or in the jargon of the television business, refuse to "clear." Edward R. Murrow was not granted clearance by over half the CBS affiliates. At one time 23 ABC stations were not clearing for the ABC News. ABC also had trouble in 1977 when affiliates declined to clear for *Soap* until Fred Silverman did some cajoling. Such events are exceptional, though. Affiliates need the glittering programs of the networks, and the audiences they collect, if advertising spots are to be sold. It is not in an affiliate's long-term interest to get a reputation as a station that refuses clearance.

One way the networks themselves can influence content is through their in-house censors. These are employees who try to make sure that everything is removed from a show or commercial that might be offensive or inaccurate. At CBS this activity goes on in the Office of Program Practices. About 70 people are involved in this nit-picking, looking for anything that might irritate the public, the Federal Communications Commission, the Federal Trade Commission, the National Association of Broadcasters, CBS itself, or any other group or person. During the course of a year the Office will scrutinize 2,000 prime time scripts, 1,500 game shows, 1,300 soap operas, 200 feature films, 200 Saturday morning shows, 33,000 commercials. The commercials are the worst offenders—CBS will send about a third of them back to be corrected. No one in the networks wants a misled and irate audience if it can be avoided.

In theory critics can also have an effect on programming, but in practice they don't. "The main effect of television criticism," says Hugh McCarney, "is to anger a few television executives." The reason that most television critics have been reduced to writing up previews and profiles is that they have no authority with the public, since the public is not of their critical persuasion. "Dismaying as it might be to a TV critic, the response of the American public over the last two decades has constituted an overwhelming endorsement of television just the way it is," writes *Wall Street Journal* television reporter James MacGregor.

We have come back to the audience again, and back to the realization that television content is shaped almost entirely by the wants of the viewers. Other influences upon programming are small compared to the likes and dislikes of the American audience, spotted and relayed by the A. C. Nielsen Company. These desires are what broadcasters are determined to satisfy. Their success at doing this was demonstrated one more time by a 1981 Roper poll in which a sample of 2,000 adults was asked what in life gave them the most personal satisfaction, day in day out. Television came in second only to family, and outstripped friends, music, reading, home, work, eating.

It is difficult to say without flinching, but it is absolutely the case that television is watching us. We watch it, and by means of the Nielsen ratings it watches us in return. The more it sees of us, the more it learns about us, the better it serves us.

And Yet . . .

It is all very simple. The economics of the television business compel each network to broadcast shows it hopes will be the most alluring. As viewers we cluster around those fantasies we find to be the best at mental cleansing and redemption. A network is not out to help anyone except itself, but to do that it has to transmit programming that turns out to be therapeutic.

And yet, the argument that television is good for people meets great resistance. It can be hard to swallow the proposition that those clichéd, paper-board productions are of great psychological value. Most of us are at times susceptible to the anti-television sentiments which in the extreme I'm calling Media Snobbery.

Part of the reason people so readily slam television is that tolerance of the lower reaches of the mind remains limited. We intuit that we can't afford to give the unconscious much play if we want to be successful in the real world where conscious behavior reigns. Even though the unconscious reaches out for televised fantasies, and makes good use of them, still at the level of mental activity where rationality prevails, stern judgments are passed. It is a form of psychic snobbery prevalent in Freud's day and not extinguished in ours: the conscious mind disdains the unconscious.

But the larger part of Media Snobbery stems from social, not psychological snobbery. The affluent and the highly educated, it has been demonstrated, watch virtually as much, and virtually the same sorts of programming as everyone else, yet they strongly denounce the medium and its uncritical viewers. And many other Americans, not wanting to be among the outs in society, rush to join with the Snobs. Media Snobbery is an ideology which is probably hypocritical and certainly imperious.

5
Media Snobbery

Nicholas Johnson

During his years as a Commissioner on the Federal Communications Commission, from 1966 to 1973, Nicholas Johnson was the most prominent Media Snob in America. He perceived television as "one of the most powerful forces man has ever unleashed upon himself," and he was determined to do battle. What was wrong with the medium? In a 1970 interview with Mitchell Kraus, Johnson talked on about "a most serious subject in our country today, and that is what commercial television is doing to mess up our heads in the way we perceive ourselves, and the world about us, and our lives, and preaching at us constantly standards of conspicuous consumption, and hedonism as the sole salvation; a sense of one's worth as an individual to be measured by the number of products he buys; the suggestion that all of life's problems can be immediately dispensed by taking a chemical into the body, or spraying one on the outside of the body." Television was deeply immoral in the way it goaded people into spurious views and unnecessary purchases. The manipulation of humans was most sinisterly done in the case of children, he held. Many people were distressed about television for youngsters, but none of them phrased it quite as Johnson did. The networks, he preached in 1972, have "molested the minds of the nation's children."

It was odd that Nicholas Johnson ever came to be an FCC Commissioner, for customarily (with the possible exception of Newton Minow) these offices have gone to people with ties to the communications industries—people with some promise for mediating between the goals of high-minded legislation and the realities of the broadcasting business. Until Johnson joined them, the Commissioners had never turned down a license renewal application. In the habit of acting more as trustees than regulators, they wanted to see that everything went agreeably, something Johnson did not care about. Sparks flew. An unnamed Commissioner was later quoted in *Broadcasting* as recollecting about Johnson, "His positions were so extreme, so vitriolic, that he lost the confidence of his colleagues. Whatever he brought up was looked upon with distrust and suspicion. I didn't trust him. I did not think his intent was to find constructive solutions." Another Commissioner, Kenneth Cox, reported on Johnson's explanation for being a publicity hound: "Nick said that the minority on the FCC had always carried on their battle in a closed manner, writing dissents that are listened to only by broadcasters. He said he wanted to break out of that. He wanted to create a public base."

Nicholas Johnson's appointment to the FCC can be seen as a whim of Lyndon Johnson's. Although not a Texan (he grew up as the only child of an Iowa speech professor), the young Johnson attended the University of Texas, graduated Phi Beta Kappa, and went on to law school there. He came to the President-to-be's attention when he served as a law clerk to Supreme Court Justice Hugo Black. When LBJ had to fill the position of Maritime Administrator in 1964, he tapped the up-and-coming University of Texas graduate, making him at the age of 29 the youngest Administrator ever. It took Nicholas Johnson very little time to reveal a penchant for stirring up dissension. By meddling with federal subsidies for shipbuilders and maritime unions, Johnson precipitated an outcry from the industry which was long and loud. The only way for LBJ to make peace was to transfer the Administrator. An opportunity appeared in the form of an open seat on the FCC, and there the young lawyer was sent in 1966.

Once at the FCC Johnson began his assault on television. He restated the opinions of those experts who felt that the medium was at fault for the violence in America. Speaking at the 1969 hearings of the National Commission on the Causes and Prevention of Violence, he quoted Albert Bandura (the researcher behind the famous Bobo doll experiments): "It has been shown that if people are exposed to televised aggression they learn aggressive patterns of behavior. There is no longer any need to equivocate." Johnson's speeches and articles were bound together in a book—*How to Talk Back to Your Television Set* was the title of it—and in it are found Johnson's own words on television and violence. "One cannot understand violence in America," he wrote, "without understanding the effects of television violence upon that violence."

It was not just violence that was traced to television in *How to Talk Back to Your Television Set*. It was many of America's social ills. "How many more crises must we undergo before we begin to understand the impact of television upon *all* the attitudes and events in our society?" Johnson pleaded. "How many more such crises can America withstand and survive as a nation united? Are we going to have to wait for dramatic upturns in the number and rates of high school dropouts, broken families, disintegrating universities, illegitimate children, mental illness, crime, alienated blacks and young people, alcoholism, suicide rates and drug consumption?"

Such failings, in Johnson's perspective, were due to the outsized, profiteering television industry. Its greed led it to beam numbing, belittling messages at the audience—a state of affairs that continued because of the lack of proper controls on the media. The networks were being allowed to undermine all that was good and constructive in American life. "What right has television to tear down every night what the American people are spending $52 billion a year to build up through their school system?" he implored.

But the situation was not beyond repair. Johnson envisioned a much altered media system, in which television would be charged with carrying out nobler functions. The waywardness of the past little while could be corrected. "The media must mold the opinion of tomorrow's polls—not, like the calculating candidate, simply mirror the passion of yesterday's mobs. They must educate," he urged.

This didactic role for television would receive much guidance from a proposed "Citizens' Committee on Broadcasting," composed of between 50 and 200 specially selected experts. As Johnson imagined it, the

Committee would operate apart from both the government and the broadcasting industry. It would monitor broadcasting and investigate prevailing practices. The vigilance of the Committee would result in program standards which would help shape the content of television.

Whether or not this is a potentially dangerous proposal, it is one that might be expected from a Media Snob. A Media Snob is someone who is scornful about television, and who refuses to understand how the medium presently works and what the benefits are that it brings its enormous audience. A Media Snob is also capable of being condescending about people who do watch the shows. As mentioned before, there is a bit of Media Snobbery in everyone, and from time to time it rises to the surface when we chastise television programming. But we should be aware of where these sentiments often come from and what they truly signify.

The Roots of Media Snobbery

For years Media Snobbery has been the most spelled-out response to the coming of television. Considering the force with which television plowed into American life, it is not surprising that this response has been largely reactive and negative. If Americans were bowled over, Media Snobs were at least trying to get to their feet and fight back.

Nor, as we'll see, is it surprising that Media Snobbery would issue from that stratum of society which most sensed itself to be challenged by the onrush of television. It is only reasonable that the more privileged groups would feel resentful toward an upstart medium which in their eyes was outrageously plebeian. People should not have been startled when in 1961 President Kennedy's newly appointed Chairman of the Federal Communications Commission, Newton Minow, a wealthy lawyer, lashed out at a national convention of broadcasters, "I invite you to sit down in front of your television set when your station goes on the air and stay there without a book, magazine, newspaper, profit and loss sheet or rating book to distract you—and keep your eyes glued to that set until the station signs off. I can assure you that you will observe a vast wasteland. You will see a procession of game shows, violence, audience participation shows, formula comedies about totally unbelievable families, blood and thunder, mayhem, violence, sadism, murder, western bad men, western good men, private eyes, gangsters, more violence, and cartoons." (And people probably should not have been puzzled to read some time later that Minow confessed he himself enjoyed watching television, especially one favorite show, *Get Smart.*)

Media Snobs hold two apparently contradictory opinions about the damage television is supposed to be doing. The first is that television speeds viewers up, and the second is that television slows them down. As *Washington Post* television writer Tom Shales puts it, "Some say TV has created a generation of snarling vicious dogs who rape and maim willy-nilly, and some say it has created a culture of benumbed ciphers drained completely of the will to fight back." Books by Media Snobs often put forth both positions simultaneously.

Regarding the first issue, that of violence, college professor Rose Goldsen writes in her *The Show and Tell Machine,* "Many people of good will find it hard to believe that a national menu specializing in shows centered on killing has no ill effects." The bent toward violent shows is in the very nature of the machine, states Jerry Mander in *Four Arguments for the Elimination of Television.* Since television has such poor picture quality, he argues solemnly, it is best suited to the depiction of overblown emotions of the rock-'em sock-'em sort.

In the second, contrary tenet of Media Snobbery, television does not so much stir people up as knock them out. Rose Goldsen refers ominously to "a single massive desensitization session conducted daily and nightly via coast-to-coast hookup." Jerry Mander proclaims, "Television suppresses and replaces creative human imagery, encourages mass passivity, and trains people to accept authority." Media Snobs frequently declare that viewers are being drugged into submission by the medium.

Actually, both Media Snobs and the rest of the public agree that television tranquilizes people. "Relaxed" and "restful" are what viewers say they feel after watching a few shows. There may be little agreement about the relationship between televised violence and real world violence, but on this second issue there is consensus regarding effects. The question is whether or not tranquility is good. In his book Jerry Mander relates about his own viewing: "Even if the program I'd been watching had been of some particular interest, the experience felt 'antilife,' as though I'd been drained in some way, or I'd been used. I came away feeling a kind of internal deadening, as if my whole physical being had gone dormant, the victim of a vague soft assault." He is describing in pejorative terms that same sensation that most viewers are actively seeking.

One thing that authors Goldsen and Mander share with most other Media Snobs is an adoration of the printed word and a conviction that television is undercutting it. "The literacy that goes with books and literature can free the mind, stretch the imagination, liberate the reader from his bondage to the present, linking him back to all of human history, all of human culture, all of human experience," proclaims Rose Goldsen, but: "Television now holds a virtual monopoly on whatever artistic and symbolic forms have a chance to be widely shared throughout the society." Mander agrees that print is the preferable means of communication, and maintains, "With books you are at least able to stop and think about what you read. This gives you some chance to analyze. With television the images just come."

This is as good a point as any to begin countering the notions of Media Snobs. Their comparison of print media and electronic media is a jaundiced one. The truth of the matter is that as television-viewing has increased over the last three decades, so has book-reading. There are more members of book clubs, higher circulation of library volumes, a larger number of books sold per capita. Americans are reading as never before.

It is misleading to suggest, as Mander does, that a book-reader is in control of his situation while a television-viewer is not. Sets can be turned off as easily as books can be put down. But more generally, people watching television regulate their intake by simply letting their attention wander or by starting up another activity. John P. Robinson, the social scientist who established how Americans use the hours available to them, reports that for about half the minutes spent with television people confess that they are also doing something else—visiting, cleaning, eating, dealing with children. The viewer, not the set, controls the process. If anything it is probably books, coming with the cachet of authority, which are the better candidate for the accusation of forcing their way into people's brains.

When Snobs compare books and television, they frequently are not comparing typical examples of each. For them all books may be symbolized by the exceptional one which illuminates truth and beauty through the eloquence of its prose. A Media Snob, in the words of one thinker on popular arts, Gilbert Seldes, "talks about 'the art of fiction' or 'the dramatic art' as if 90 percent of the books and plays offered to us each year weren't unmitigated trash." Conversely, all of television is supposed to be equivalent to the shoddiest content ever sent over the airwaves. Instead of upholding these stereotypes, if Snobs were to take an average book—a mediocre piece of detective fiction, for instance—and compare that to middle-range television, the qualitative differences might even out.

Why do Media Snobs insist that the printed word is succumbing to the televised image? A hint comes from Rose Goldsen, who writes ingenuously, "I confess I find it frightening to see control of access routes pass out of our own hands, out of the hands of artists, craftsmen, and lovers of art whose primary allegiance is to a work's authenticity, and into the hands of a small group of anonymous men and women ill equipped to take charge, unaware even of the extent of their responsibility." She is dismayed that management of communication channels is no longer exclusively in the hands of her sort of person, but now must be shared with others alien to her. This is a waspish but very human sort of reaction, experienced to some extent by everyone who is forced to shove over and make room, no matter what the context. The unfortunate thing is that in this instance a whole set of outlooks, an entire philosophy, has been constructed by Media Snobs on the base of what seem to be peevish feelings.

There is more to the heated reaction of Media Snobbery than the competition between an older and a newer mode of communication. Partisans of these two modes are found at very different layers of the social hierarchy. Media Snobs are at home in the small patrician strata at the crest of society; literacy and literature are the markers of their membership. Naturally they are reluctant to let go of their favored position, and so set themselves in opposition to the social changes which they feel are benefiting the larger number of people at their expense. Because the popular culture which television carries is emblematic of these detested changes, it catches the brunt of their resentment. At bottom, Media Snobbery can be antidemocratic, embraced by those who dream of old modes of life where their social station was more hallowed and certain.

The underlying issues in the debate over television are those of social power. Sociologist Herbert Gans of Columbia University has observed that anti-television sentiments originate in class antagonisms, and in aristocratic longings and attitudes which have yet to die out. What here is being called Media Snobbery, Dr. Gans concurs in his *Popular Culture and High Culture,* is "a plea for the restoration of an elitist order by the creators of high culture, the literary critics and essayists who support them, and a number of social critics—including some sociologists—who are unhappy with the tendencies toward cultural democracy that exist in every modern society."

In Gans's analysis, those who make the anti-television critique today belong to a class that a few centuries ago dominated cultural life. These were "the city-dwelling elites—the court, the nobility, the priesthood, and merchants—who had the time, education, and resources for entertainment and art." With the advent of industrialization, changes in employment and the distribution of wealth brought about the rise of a huge market for the popular arts. Threatened and resentful, the precursors of Media Snobs were "fearful of the power of popular culture, rejected the desirability of cultural democracy, and felt impelled to defend high culture against what they deemed to be a serious threat from popular culture, the industries that provide it, and its publics."

Just as Gans implies, history reveals there is little new about Media Snobbery. The basic beliefs of a Media Snob are not original in the television era. They have been around as long as there has been a popular medium which was perceived to challenge the established social order. As soon as the patricians felt their position weakening, they fought back by deriding the culture of the plebeians. In England this began during the 18th century when popular literature was first appearing. In the United States in the 19th century it found its target in the penny paper and the dime novel, and in the first half of the 20th century it was radio and the movies. A minister wrote in 1919:

> The tendency of children to imitate the daring deeds seen upon the screen has been illustrated in nearly every court in the land. Train wrecks, robberies, murders, thefts, runaways, and other forms of juvenile delinquency have been traced to some particular film. The imitation is not confined to young boys and girls but extends even through adolescence and to adults.

And a prominent critic in 1930:

> The movies are so occupied with crime and sex and are so saturating the minds of children the world over with social sewage that they have become a menace to the mental and moral life of the coming generation.

Television has simply become the focal point for a traditional disparagement. The less stratified society becomes, and the larger the middle class grows, then the more vitriolic the dislodged elite wax.

If Snobs cannot be outrightly hostile to the growing majority, they can be condescending. Snobs are liable to look down their noses at people who freely enjoy what television brings. Studies confirm that those who have reached high educational levels are the most critical of television, and the most likely to believe it's intended only for the less educated. Snobs will say patronizingly that television shows are designed for "the lowest common denominator." The use of this term, in fact, is one thing that marks Snobs off from everyone else. Seventeen years after he had described television as a "vast wasteland," on the day that he became chairman of the Public Broadcasting Service, Newton Minow was asked how commercial television had changed over the interim. "There's not been much improvement," he replied, "particularly on the entertainment side. There has tended to be an almost regular sinking to the lowest common denominator."

Let's consider this term, "lowest common denominator." In mathematics, where it came from, it is positively conceived of, and something to aim for. For a group of fractions, it is the numerically lowest base which they all can be converted to, so that calculations can go on readily. Snatched from this application and applied to society, its implications change radically from positive to negative. The words "low" and "common" become derogatory, at least for a class-conscious Media Snob. Shows for the "lowest common denominator" are supposed to be churlish and tasteless, although in truth they are nothing more or less than shows for the majority. There is nothing wrong with "the lowest common denominator" in mathematics, or in an egalitarian land. Former FCC Commissioner Leo Loevinger has dealt thoughtfully with this term: "The cultural denominator of popular programs may be the *highest,* not the lowest that is truly common. The important point is that as television lets us share daily a common reflection of society and helps us see a similar vision of our relationship to society, it builds a common culture to unite our country. This appears to be its natural function and highest ideal."

As much as Snobs may publicly belittle television viewers, it appears to be the case that privately they watch as much video fantasy as anyone else. George Comstock, chief author of the compendium *Television and Human Behavior,* summarizes the 1960 Steiner survey and the 1970

Bower one when he writes, "Despite the more frequent declarations of a desire for more informational and educational programming, the television diet of the better educated was about as heavily weighted with entertainment as that of the less educated." More explicitly, Bower had constructed what he called a Culture Index, calibrated by such activities as listening to operas and going to ballets. When he compared the television habits of those high in the Index with those at the low end, he discovered that the highs spend 25 percent of their viewing time with comedies, while those at the low end spent 26 percent. Time with action/adventure shows was equally close: 17 percent for those high in the Culture Index versus 16 percent for the lows.

Whatever differences in viewing time and choices there once may have been between those of higher and lower status, the gap is closing. This is Comstock's conclusion as he looked back over thirty years of studies on audience behavior. The viewing habits of the less well-off had begun earlier and crested sooner, but the elites' time with television has continued to rise toward parity. By the mid-1970s the average viewing time of the affluent had approached the figures for the rest of society.

Every minute that Media Snobs spend ingesting television undercuts the idea they wish to spread about the poisonous effects of viewing, for they are not converted into ogres or reduced to being slugs by the experience, any more than anyone else is. If they truly believed that rousing or sedating content were a bad thing, they would have more than enough to attack within their own high culture. But operas do not get berated for raising passions, nor symphonies for soothing them; novels are not criticized for their turbulent action, nor poetry for its calming rhapsodizing. It is only when the greater number of people experience the same results through the popular culture of television that Media Snobs sally forth. Their barrage of criticism is loosened not to stave off the downfall of their fellowman or for any other glorious purpose but to save themselves from being swamped by the rapid expansion of the middle class and its culture.

Misconceptions

As Media Snobs, Nicholas Johnson, Rose Goldsen, and Jerry Mander have much in common. They each identify closely with the loftiest traditions of education and literacy, and they each are repulsed by the swift incursion of television. They also share in two deeply flawed conceptions about the nature of the new medium: they think that messages flow only one way in the television system, from networks to viewers; and they think that what television delivers is instruction.

Media Snobs close their eyes to the fact that the television system is circular, and that messages are also sent back from the audience to the broadcasters. The importance of the feedback that comes via the ratings is not something that Snobs care to admit. They find it convenient to ignore that the senders and receivers of television programming are very much in tune with each other, just as senders and receivers are in any successful communications situation. Snobs would be right to infer that the networks and the viewers were not linked together by a two-way flow of communication if the audience turned away to paperbacks, movies, radio shows, live sports, comic books, or other fantasy sources, and if the television system collapsed as a result. But since the system has endured, then the feedback which brings stability must be pumping through it.

It is a one-way model of mass communication which underlies the thinking of Media Snobs. Television forces itself upon people whether they like it or not, Snobs suggest. This conception of television as bully is implied in the title of Johnson's book, *How to Talk Back to Your Television Set.* He is pretending that the public does not presently respond to the broadcast industry, and is not the most picky participant in the process of television programming. Jerry Mander refers to television as "the most powerful mind-implanting instrument in history," and Rose Goldsen is even more graphic when she calls the medium a "cattle prod." Snobs misperceive the loop as a lance.

The lance-throwers, in Media Snobbery's version of the mass media, are monolithic broadcasters. The television industry is conceived of in the way that giants are described in myths and fairy tales—all-powerful, towering, up to some horrendous business. Johnson described it as "without question the single most economically and politically powerful industry in our nation's history." Rose Goldsen too views it as oozing power: "The power to dominate a culture's symbol-producing apparatus is the power to create the ambiance that forms consciousness itself. It is a power we see exercised daily by the television business as it penetrates virtually every home with the most massive continuing spectacle human history has ever known."

In *Understanding Media* Marshall McLuhan came closer to the truth when he subtitled his chapter on television "The Timid Giant." Timidity does characterize the industry, for it is terrified of doing anything that might cause the audience to waver in its affection. As to whether or not it is a giant, television does not appear gigantic when stood up alongside other industries in American life. It is smaller in financial size than the tobacco industry, or the antiques business. Much of the giganticism a Snob sees in television is in the eye of the beholder.

The viewers in Media Snobbery's one-way model of television are thought of as if they were staked-out victims of the broadcasters' lances. Into their heads can be drilled all matter of content, according to this uncomplimentary view. Television images penetrate Americans' brains and refashion them according to its own liking, says Rose Goldsen: "It is minds they make—and minds are always in the making." Jerry Mander agrees that Americans have succumbed to television, crying, "We have lost control of our images. We have lost control of our minds." Never mind that the likes and dislikes of the public determine programming, and that very few of the networks' hopeful offerings will pass muster with the fussy audience; as Snobs tell of it, it's viewers who are the supplicants.

Why do Media Snobs insist upon a simple-minded rendition of the reciprocities involved in mass communication? Why are they reluctant to acknowledge that the content being broadcast is pretty much the content the audience is ordering up? There are several possible reasons.

Media Snobs' mistaken view of the media may derive from their outmoded sense of the nature of social life. Worshipping the past, they appreciate the world as it was, and insist that's how it still is. They are inclined to perceive a society that is more stratified than America's is today, more rigid, more governed by conventions of dominance and deference. Having an authoritarian's perspective on things, seeing the world in terms of higher and lower ranks, they peer into the dimly visible mechanisms of mass communication and manage to find there the same sort of pattern, in which a looming television industry beams images at a hapless audience. Inferiors are under the control of superiors, just as in their reveries. Their antiquated model of the world—valid decades ago, yet still saluted by Snobs—is a deficient model of how communication works today.

It is possible there are psychological factors bound up in Snobs' choice of the one-way model. Their particular mental image of television might be telling about the nature of their own deepest feelings. When Snobs look at a communications situation which they are unable or unwilling to understand, one as vague to them as an ink blot, they may ascribe to it sentiments which lie deep within themselves. If they sense subconsciously they are being edged out of their rightful place in society, they could be experiencing strong and vengeful emotions. They may insist that television is doing what they themselves secretly want to be doing—gunning down the common man, blowing the majority to smithereens.

If Media Snobs were to concede that television is best described as two-way and transactional, then they would have to concede that the system was sound in that the senders and receivers were attuned to each other and communication was taking place. This is an admission Snobs would be loath to make, for their interests are best served by an insistence upon a broken, discreditable system. In the teeth of all the evidence to the contrary Nicholas Johnson feels compelled to assert, "To say that current programming is what the audience 'wants' in any meaningful sense is either pure doubletalk or unbelievable naivete." He's the one guilty of doubletalk and naivete here, but if he reversed himself and admitted that current programming is what the public wants, his condemnation of television would be robbed of much of its indignation.

The second misconception of Media Snobs dovetails with the first. Not only do Snobs blindly misperceive the television system, they also misunderstand its content. The messages which would best suit Media Snobbery's lame model of mass communication would be what Gerhardt Wiebe called *directive* messages—those which impart new information and call for learning and adjustments. Such messages depend more on the force of the transmission than on the willing reception of the audience. And these are just the messages Snobs contend television carries, for they believe the medium's prime effect is to teach. They talk as if television trafficked in information or instruction for the greater part, not in fantasies.

"Consider what we learn about life from television. Watch for yourself, and draw your own conclusions," urges Nicholas Johnson, proposing a kind of intense viewing done with all one's critical and conscious faculties at the fore. But people don't watch television that way, and so they don't generally learn from the medium. At another point Johnson remarks, "By the time the average child enters kindergarten he has already spent more hours learning about his world from television than the hours he would spend in a college classroom earning a B.A. degree," in the belief that these are the same kind of experience, instead of the opposite. Compounding his error, Johnson wants television to stop doing what it's doing and start up the brand of instruction he favors, as if the medium could.

Rose Goldsen as well believes television is instructive, saying vividly that "the United States enjoys the dubious distinction of having allowed the television business to score a first in human history: the first undertaking in mass behavior modification by coast-to-coast and intercontinental electronic hookup." As Jerry Mander sees it, "the viewer is little more than a vessel of reception," and television "trains people to accept authority."

However, television could hardly be worse at putting information into brains or causing viewers to change the way they are. Considering the enormous size of the audience, and the extraordinary number of hours spent viewing, the most astounding feature is that so little is absorbed. Only a very few of the countless images which television sends the audience's way stick in brains; all the rest come and go, rippling on. Television pulses through the unconscious, cleaning it out, and has remarkably little effect upon the conscious mind or upon memory. A study on viewer comprehension was done in 1980 for the American Association of Advertising Agencies, and the results could only be surprising to someone who didn't understand the functions of the medium. More than 90 percent of viewers misperceive at least part of whatever kind of programming they watch. People routinely misinterpret between one-fourth and one-third of any broadcast, whether it's entertainment, news, or commercials. Would the figures have been even higher if the study hadn't been commissioned by an organization with a vested interest in maintaining that television can teach? Perhaps. For television can't, or more precisely, can't very much.

In the modern era, instruction continues to come from where it always has—the real world. The family, the schoolroom, and the workplace remain the touchstones of Americans. In those situations people learn the sharp lessons in how the world works and in what its tolerances and protocols are. There values are hammered out, attitudes are molded, abilities are honed, ambitions are made reasonable. The potential losses are too disastrous and the potential rewards are too attractive to allow for much misperception and miscalculation. Every functioning human is a person who has been trained and retrained by reality, and who has brought his behavior within the range of the permissible and directed it toward the praiseworthy.

It is true that in the absence of real-world information television can offer some hunches. But for most people these are only provisional and will be discarded whenever they are controverted. For instance, many people do not have close contact with the professional roles that television drama makes full use of. When we actually come to deal with a doctor or lawyer or policeman, our expectations may be shaped by the behavior of video heroes. Then the doctor may be prompted to explain that he is not Marcus Welby, or the lawyer that he is not Perry Mason. When the policeman scoffs at Kojak, most people will adjust to the newly apparent reality. For it is reality which is the binding lesson. Fantasy does not override real-world familiarity. No high school student expects to enter *Room 222* or be taught by Gabe Kaplan.

In a chapter entitled "The Myth of a Lack of Impact," Nicholas Johnson tries to deny the fact that television does little by way of instruction. One of his counterarguments is that Dr. George Gerbner had proven it does. By far the most conspicuous social scientist digging into the question of television's effects, Gerbner is dean of the prestigious Annenberg School of Communication at the University of Pennsylvania. Together with his colleague Larry Gross, he has been keeping close track of televised violence since the 1967–68 viewing season. Annually in the spring Gerbner and Gross release their "Violence Profile," which purports to reveal violence levels for the season just past. The figures are widely disseminated by the wire services, and appear in local newspapers commonly under a headline stating the percentage increase or decrease from the previous year.

Along with the Violence Profile comes what Gerbner and Gross call their "Cultivation Analysis," which supposedly measures the social effects of television mayhem. The theory of Gerbner and Gross is that broadcast violence is accepted by certain segments of the public as information about the real world. The two researchers state that the medium "cultivates" fright and anxiety in these wide-eyed viewers. "The prevailing message of television is to generate fear," they say. For proof they claim to have found a statistical correlation between heavy viewing (four or more hours per day) and exaggerated perceptions of threat (gauged by asking people to estimate the chances of being involved in a violent incident, and then comparing the guesses to known national figures). This is the kind of instruction Nicholas Johnson is referring to.

Assuming for the moment that a correlation does exist between heavy viewing and fearfulness, it is not clear that Gerbner and Gross's theory would provide the best explanation. Recall that, according to Bower's shrewd analysis, heavy viewers are those with the greatest opportunity to view—they are not in the labor force. These are women, the poor, and the elderly—in short, the powerless, and the most likely to actually be victims in the real world. It could be the harsh realities of their lives and communities, and not the unrealities of television fantasy, that lead them to predict high chances of violence. Truth, not misguidance, could be behind Gerbner's figures. And in fact, another study which made allowance for the crime rates in respondents' neighborhoods could find no statistical relationship between the extent of television-viewing and fear of being a victim.

But theories aside, it turns out that Gerbner and Gross's demonstration of a correlation between viewing and apprehension is deeply marred. An examination of their statistics was published in the journal *Communication Research* in 1981 by Dr. Paul Hirsch, a sociologist at the University of Chicago. Hirsch said he found it amazing that work as important as that of Gerbner's, which had exerted such a pronounced influence on thinking about the mass media, had undergone so little scrutiny by other social scientists. A 1978 study had discovered that Gerbner's Cultivation Analysis did not apply to British television or viewers, but no one in the United States had troubled to reanalyze Gerbner's original data. This Hirsch did.

What Hirsch found was at variance with Gerbner's conclusions. Instead of heaviest viewers being the most fearful, Hirsch learned that *non-viewers* were the most frightened of all. And that within the category of heavy viewers, those who watch eight or more hours are *less* fearful, not more, than those who watch four to seven hours daily. The relationship between viewing and fearfulness was highly inconsistent. Hirsch also discovered that the apprehension of people in such victimized groups as blacks, females, and the elderly was statistically independent of the amount of television seen. All in all, Hirsch ended with scholarly reserve, "acceptance of the cultivation hypothesis as anything more than an interesting but unsupported speculation is premature and unwarranted at this time."

Another leg is pulled out from under the Snobbish conviction that television puts things into brains. I am not trying to say that television does not teach anything, for clearly it does to some limited extent. Television news does bring the public a dollop of information, as we shall see in a later chapter. At election times the medium has a great deal to tell us about the candidates. And commercials alter enough behavior to make the effort worthwhile to advertisers. Even entertainment shows can instruct: when a 71-year-old bank robber was foiled in Seattle, he remarked, "It looked easy on TV." Although less reported, the shows occasionally teach positive lessons too: in Chicago two brothers saved the life of a comatose man they had pulled from a burning building by giving him a heart massage they had seen on *Emergency*.

But when discussing the instructive capabilities of American television, a sense of proportion is needed. For the five million hours of programming broadcast annually, to an audience of over 200 million people viewing several hours daily, the amount of learning is undeniably miniscule. The reason is that *directive* messages are not what viewers want, so they are not what networks can afford to send. Snobs are deceiving themselves in thinking television steers people around. The audience demands what Wiebe calls *restorative* content—those slight fantasies that sponge minds free of tension.

6
Television Is Good for Your Nerves

Situation Comedies

In the fantasy diets of American viewers the main dish, the one that sits in the center and gets dipped into by everyone, is situation comedy. Other varieties of programs draw their fans unequally from the standard demographic categories, but comedies are relished by the young as well as the old, males as well as females, rural as well as urban citizens. Universal favorites, they dominate the top forty slots in the Nielsen ratings, rising from 30 percent in 1970 to 45 percent in 1980. People are willing to watch the same episode time and again, making situation comedies the kind of serial most frequently resurrected through syndication. The nation's desire for this sort of broadcast fantasy appears near-insatiable.

The networks do their best to keep offering after offering coming

THE ELMER E. RASMUSON LIBRARY
UNIVERSITY OF ALASKA

along, sometimes reaching far down into the ranks of available writing, acting, and production talent. No other medium delivers comedy to the extent that television does; it is a segment of movies, novels, magazines, and theatre, but it is not the core. "Comedy," wrote Gilbert Seldes in his book *The Popular Arts,* "is the axis on which broadcasting revolves."

Situation comedy is the easy favorite of everyone involved—as well as viewers and broadcasters, the unctuous sponsors too. A successful show, collecting a gargantuan crowd of beaming solvent citizens, could hardly be a more congenial occasion for the half-dozen commercial pitches. And while situation comedies are not altogether free of the kinds of controversies that start advertisers twitching, as a generality they have fewer trouble-spots. No wonder advertisers stand in line to buy the thirty-second niches on the programs to set out their wares.

Within the world of situation comedies there is ample range in tastes, from *M*A*S*H* and *Barney Miller* at one pole to *Hello, Larry* and *Three's Company* at another. The robust variety augurs that situation comedy will continue to engage the American audience far into the future. Jeff Greenfield notes, "When three networks try to tickle funny-bones in 76.3 million households, there is no one style or level they can rely on. So there is no special kind of comedy we'll be watching in the years to come—just a lot of it."

A reason to be sure about the projected prevalence of situation comedies is that the past trend has been so strong and unwavering. Demand for these shows has never slackened since the beginning of television broadcasting. The half hours of humor got off to a flying start with the audience of the '50s because the format had already been well engineered in radio days, and finely adjusted to viewers' likes and dislikes. With a minimum of difficulty the radio serials transformed themselves into television serials. *Amos and Andy,* the very first radio comedy show, became one of the first television comedies, to be followed by *Our Miss Brooks, The Adventures of Ozzie and Harriet,* and dozens of others. The more the new medium penetrated the American public, the more the call went up for situation comedy.

The honor of being the most popular of the early situation comedies belongs to a show that can still be seen in black-and-white reruns almost everywhere, *I Love Lucy.* Lucy had also shifted over from radio, where her show *My Favorite Husband* had been popular during the 40s. When *I Love Lucy* was first aired on Monday, October 15, 1951, it was with some misgivings on the part of the sponsor, Philip Morris. Company officials felt that a comedy featuring a stagestruck redhead and a Cuban bandleader was doubtful fare for the American audience. But the show was immediately and spectacularly successful. It was the top-ranked program of the year, and the first one ever to be seen by an audience of over 10 million households. On Monday evenings most other activity in the United States came to a stop. Marshall Field department store in Chicago was forced to display a sign in its window, "We love Lucy too, so we're closing on Monday nights." During its nine-year run the show won over 200 awards, including five Emmys. Never out of syndication, its 179 episodes are now available in Spanish, French, Italian, Portuguese, and Japanese as well as English.

The success of *I Love Lucy* was in part due to the considerable talent of the two principals. Lucille Ball is one of the funniest comediennes ever to mug before a camera; no one who has seen her do low-comedy skits can shake the memory. At the close of the first season *Time* magazine lauded the actress: "What televiewers see on their screens is the sort of cheerful rowdiness that has been rare in the U.S. since the days of the silent movies' Keystone Comedies. Lucille submits enthusiastically to being hit with pies; she falls over furniture, gets locked in home freezers, is chased by knife-wielding fanatics. Tricked out as a ballerina or a Hindu maharanee or a toothless hillbilly, she takes her assorted lumps and pratfalls with unflagging zest and good humor. Her mobile, rubbery face reflects a limitless variety of emotions, from maniacal pleasure to sepulchral gloom. Even on a flickering, pallid TV screen, her wide-set saucer eyes beam with the massed candlepower of a lighthouse on a dark night."

Lucille's abilities were complemented by those of her husband, Desi Arnaz, who developed into a perfect foil for her antics. Desi served well as a slightly ludicrous authority figure, hamstrung by a broad Spanish accent. It was against his dictums that Lucy was constantly bumping. He may have come out the victor—the very first episode began with Lucy wanting to go to a nightclub, and ended 22 minutes later as they went to the boxing matches—but it is Lucy and the audience who are the real winners.

Beyond being a straight man, Desi's greater contribution was behind the scenes, operating the production company. *I Love Lucy* was the end product of a highly complex and tightly managed enterprise, and Desi was in charge of it all. The manner in which each episode was actually shot was his brainchild. He had wanted a live audience for the gusto it can bring to a performance, but he had also wanted to use film because he anticipated even at the outset that the program would be good enough to be resold. These seemed incompatible: no studio audience could sit good-heartedly through the usual hour upon hour of film production. Desi devised a system where three or more cameras were used simultaneously, scenes were shot just once, pauses were taken only to set up new scenes, and the entire show was filmed within sixty minutes. It has proved to be a method hard to supersede, and decades later is still in use for the production of situation comedies.

The talents of Lucy and Desi were one side of the phenomenal triumph of *I Love Lucy*; on the other side was the thirst of the audience. For all its pleasures and benefits, the revived family life of the '50s also generated new tensions for Americans, and it was to these discomforts that *I Love Lucy* was addressed. The focus of the show was announced in its theme song:

> I love Lucy and she loves me
> We're as happy as two can be.
> Sometimes we quarrel but then
> How we love making up again.

As quarrels flared and making up grew harder to do, Americans turned to *Lucy* for relief. Jack Gould, the then television critic of the *New York Times,* explained about the show, "Its distinction lies in its skillful presentation of the basic element of familiarity. If there is one universal theme that knows no age limitations and is recognizable to young and old, it is the institution of marriage—and more particularly the day-to-day trials of husband and wife. It is this single story line above all others with which the audience can most readily identify itself." In *I Love Lucy,* Americans of the 1950s were finding the tonic for what was bothering them.

Recollecting this serial, we come upon the first glimmerings of why situation comedy, then and now, has been in such high demand. The best situation comedies get directly at the prevailing forms of psychic distress. They deliver the fantasies which alleviate the deep mental problems of the audience. Redress is what situation comedies provide.

How do they accomplish this? First they have to get the audience to enter into their world. People must surrender to their bidding to join in emotionally. There can be no impediments between the imagination of the viewer and Lucy's apartment, or the Bunkers' living room, or the M*A*S*H compound.

The way a situation comedy achieves the fusion of its fantasy-world and the viewers' is through the use of a serial format and the sense of welcoming familiarity that it creates. With reasonable regularity the show will be there on the television screen every week at the same hour. If the program appeared just once, or randomly, or in telescoping lengths, then viewers would have to be on their toes. Their attentiveness wouldn't contribute to the business of releasing personal tension. But the serial format encouraged people to settle into a routine, and to receive the broadcasts in the most relaxed frame of mind possible. Mondays at eight, Lucy and Desi would appear. Saturdays at nine it was time for *The Mary Tyler Moore Show.*

And every week there comes the same family-like group of characters, willing to spread their comradely rapport over each viewer. No situation comedy can last for long if it fixes on just one person and omits the continuing characters that make up the mandatory human band. A familiar cluster of people, one of rock-hard durability, is the social universe of situation comedy. The same characters must have the same relationships in which they behave as they always have behaved, for as long as the show goes uncancelled. Greenfield says, "Predictability is precisely the reason for situation-comedy success. For these shows, virtually without exception, embody the central premise of American television programming: they give us characters whose habits, foibles, and responses to situations we know as we know those of our own friends and family. What's more, these characters—unlike real people—do not deviate from their habits. They provide a sense of family warmth without confusion, without ambiguity."

Lucy and Ricky and Fred and Ethel formed the mutual admiration pack of *I Love Lucy*. As close neighbors and good friends, they defined the required warm milieu. Fred might be grumpy and Ethel might be flighty (and they might hate each other offstage, as in fact they did), but they exuded loyalty, and Lucy and Ricky exuded it back. Anthropologists reporting on primitive tribes refer to the "human press"—the close gathering of the band at the end of the day as the members squeeze up against each other around the campfire. Americans by the tens of millions pant to enter imaginatively into the human press of the Ricardo's and Mertz's, or Kotter's classroom, or Barney Miller's squad room.

The human press of situation comedy is an extremely enticing fantasy for modern-day individuals. In it they can find all the plusses of real world interaction and none of the minuses. It is a place of sustaining and buoying relationships, where no one loses his way or goes friendless. Missing are thoroughgoing roguishness, repression, failure, abandonment, sickness, and death. Present are stability and warmth—the longings of the eternal child in every viewer. It is the dream of dreams that situation comedy deals in.

Once in the familiar world of a situation comedy, then viewers must be made to go along with that week's story line. Even though each episode is ostensibly different, still the underlying structure is as unvarying as a church service. Whatever happens will fall into a set, precast pattern. In the opening minutes the ongoing, amiable band will be disturbed by an unexpected development. The writers have put yet another wrinkle into the serial's nearly endless fabric: Lucy is suddenly inspired to become a ballet dancer, or Mary Richards is pursued by an unwanted beau. Complication then turns into confusion—confusion serious enough to give pause, but not so threatening that the social edifice of the neo-family is in jeopardy. The prime characters cannot lose their minds, become infirm, or move out. They have to be momentarily vexed in the story, but not unseated.

Whatever the plot brings, it must be something the audience senses can be corrected before the time is up; this is in the nature of the contract between the senders and receivers of television fantasy. Like a baby in its father's arms, viewers must know they will be flipped and tumbled, but not dropped. Week after week the plot line will stimulate their own feelings of peril, and then just as surely lay them to rest. Order will be restored, and the persevering social arrangement of the comedy reestablished. Everything is going to be all right, is the ultimate moral of situation comedy.

In a representative *I Love Lucy* segment, Lucy challenges Desi to come up with that prime emblem of marital solidarity—the date of their wedding anniversary. As millions of American husbands and wives gazed into their sets, Desi pretends to know but eludes replying. Fred and Ethel, it turns out, can't help him. In desperation he fires off a telegram to officials in the town where they were married. The arrival of the telegraphed answer is the occasion for several sight gags before Desi manages to read the correct date. Claiming he knew all along, he tells Lucy a large party has been planned and invitations have gone out. But when Lucy learns that Fred and Ethel haven't gotten an invitation, Desi is forced to confess. Things continue to degenerate until, at what should have been the celebration of the anniversary, Desi croons a love song and apologizes publicly; then the two make up. The gyrations of the plot cease and the essential balance of this situation comedy is recovered in the final minute, as always.

The term "situation comedy" is an ambiguous one because it's not clear if the word "situation" refers to the ongoing situation of the same characters sailing along in the same dramatic structure week after week, or to the kind of episodic situation like this one which momentarily rocks the boat. What "situation comedy" initially meant is unknown, since the origins of the term are obscured in the mists of early broadcasting history. No matter—the ambiguity exists because the two possible meanings are the two primary dimensions of any situation comedy. There must be the continuing situation of the familial crew within which each week's new, pesky situation is played out. These two frames give viewers the safe, resolvable fantasy world they long for after a day of ups and downs.

Once in that dual framework there's more that people want, though. This is where the "comedy" of situation comedy comes in. We are hoping for the jokes that will get those energy bursts known as laughter out of us.

Laughter

Lucy and Ricky and Ethel and Fred are sitting around after dinner. Dessert is offered and Ethel accepts, but Fred questions her. As she rises and walks in front of Fred she says, "Oh, I just want a little extra something." Fred, eyes following her, says, "From where I'm sitting you've already got a little extra something." The audience guffaws at Fred's gibe.

Jokes are the special dividends of situation comedies. One of the requisites of these programs is that they only be a half hour in length; the reason for this, supposedly, is that if they were longer the plot and characterizations would have to be more complex and the jokes would be overshadowed. *Lou Grant*'s Gene Reynolds commented, "Comedy just doesn't stretch easily into an hour because you have to get too heavy for the story and that fights comedy." Everything is done, in short, to keep the jokes in the foreground. What are jokes, and what do they do for people?

If there is one thing certain about the mysterious but universal phe-

nomenon of jokes, it is that analysis destroys them. As master writer E.B.
White once remarked, "Humor can be dissected, as a frog can, but the
thing dies in the process and the innards are discouraging to any but the
pure scientific mind." This is the first hint about the function of the jests
interspersed throughout situation comedy; they are antithetical to
rational, conscious thought. Jokes appeal to other areas of the mind.

The caveat about analysis aside, let's forge ahead. One of the very few
behavioral scientists to work up a general explanation of the nature and
role of jokes was Sigmund Freud. The title of his reflections, *Jokes and
Their Relation to the Unconscious,* gives away his thesis—that jokes
serve the mind's bottom strata. Although Freud's writing style was
highly cerebral and many of his illustrative jokes are wrecked in transla-
tion, the book is nonetheless a fine collection of insights into the peculiar
business of joking.

Freud sorted jokes into two classes—those that gave rise to only a
smile, and those that could elicit laughter. As a joke worth a smile and
nothing more he offered this example: "Experience consists in experiencing what we do not wish to experience." He noted what he claimed
was an American joke to illustrate the more substantial variety: "Two
not particularly scrupulous businessmen had succeeded, by dint of a
series of highly risky enterprises, in amassing a large fortune, and they
were now making efforts to push their way into good society. One
method, which struck them as a likely one, was to have their portraits
painted by the most celebrated and highly-paid artist in the city, whose
pictures had an immense reputation. The precious canvases were shown
for the most time at a large evening party, and the two hosts themselves
led the most influential connoisseur and art critic up to the wall upon
which the portraits were hanging side by side, to extract his admiring
judgment on them. He studied the works for a long time, and then,
shaking his head, as though there were something he had missed, pointed
to the gap between the pictures and asked quietly, 'But where's the
Savior?'" Knowing that Christ was hung between two thieves, a listener
might be moved to laugh. Out would come the merry sounds which in
Freud's view signaled the discharge of energy repressed deep in the mind.

Freud held that a joke which produced laughter and vented the
unconscious would have one of two underlying psychological purposes.
"It is either a hostile joke (serving the purpose of aggression, satire, or
defense) or an obscene joke (serving the purpose of exposure)," he
explained. The joke about the businessmen was designed to stimulate the

release of hostile feelings; other jokes exist so that repressed sexual urges and anxieties can be expelled.

Freud's conjecture about these two psychological wellsprings for jokes has been largely borne out in one of the few studies to systematically analyze televised humor. Trained observers studied prime-time television for a week, recording all incidents intended to be funny. A total of 852 attempts at humor were noted and scrutinized. Of these, 619 items, or 73 percent of the total, involved either sexual or hostile humor. Freud was about three-fourths correct when he ventured that jokes were either obscene or hostile.

It is clear that aggression of one sort or another is the crux of humor, and of television comedy. Freud said elsewhere that humor "is an invitation to common aggression and common regression." A joke encourages people to regress to more infantile, impetuous states of mind; once there, bottled up aggressions can be allowed to squirt out.

Individuals who tangle with the real world every day have to come away with some retaliatory and hostile feelings; the world presses in too much for it to be otherwise. As well-socialized creatures, humans are practiced at constructing dams in their minds to hold back their anger, for if let out directly it is largely ruinous. Also held back behind subconscious barriers are impulses bequeathed in our genetic legacy—lust, for one. Society provides inadequate outlets for aggressive energy, but joking is one acceptable means. The jokes on situation comedies permit viewers little bursts of aggression which serve to reduce mental pressure.

For a joke to do its magic it must first have established laughingstocks who are so inviting they almost tug at a person's harbored hostility. In Freud's joke the two businessmen are represented as unsavory, arrogant types, and thus are choice objects of derision. On *I Love Lucy* much of the aggressive humor is directed at women and their supposed foibles: male viewers are set to laugh at Ethel when Fred, who is no trim figure himself, pokes fun at his wife's appetite. Females in the audience, especially those concerned about their own weight, are also ready to laugh at the character presented as a dim-witted blonde.

Throughout history the two standard target areas for the symbolic blows of humor have been the status quo, and deviance from the status quo. In the first instance people counterattack against authority and the impositions it makes upon them; political leaders are made fun of, bosses and mothers-in-law are ridiculed. In the second the deviate, the oddball, the clown are the butts of the jests. *I Love Lucy* shrewdly offered up both these marks. Desi was presented as a leader—of the band and at home—who invited mockery because of his fractured English. And the carrot-topped Lucy brought ridicule upon herself for her devious, outrageous, comic shenanigans. These two targets recur in a modulated way in the characters of the gruff Fred and flighty Ethel. On *I Love Lucy,* authoritarian males and foolish females were served up to viewers as fair game for their private aggressions.

Once the laughingstocks are in position, then the punch line (a revealing term in its own right) will soon arrive. It will be something slightly incongruous and unexpected: Freud's art connoisseur invokes an unanticipated image, or Fred violates the conventional decorum of husband and wife. The punch line will trick a person and sneak by mental barricades. For a brief instant it opens a path to the unconscious mind, and the repressed energy which has been primed from the first appearance of the laughingstock now spurts out.

The aggression released by jokes departs humans in the form of laughter. When a person laughs, fifteen sets of facial muscles contract in unison. Within the torso the diaphragm bounces strenuously up and down. Air is drawn deeply into the lungs and then expelled through a series of staccato bursts. A doctor describes what happens during hearty laughter: "Out bellows the blast at seventy miles an hour. Tears are squeezed from the lacrimal glands, and there is coughing too, as the rush of air flings mucus against the lining of the windpipe, a coughing that empurples the face, engorges the veins of head and neck, and makes the words 'to die laughing' not entirely hyperbole. Shaking, writhing, weeping, coughing, you laugh. It is a violent, risky affair, a seismic dislocation, a volcano blowing off."

In their drive to elicit this discharge, and thus to increase their own popularity, situation comedies have from their earliest days in broadcasting resorted to the use of artificially augmented laughs on their sound tracks. Some situation comedies are recorded before no audience at all, and then all the chuckling and guffaws television viewers hear is fake, canned laughter. But even shows that are filmed before live audiences, as *I Love Lucy* and *The Mary Tyler Moore Show* were, still have their sound tracks "sweetened" with canned laughter before the shows are telecast, to heighten the humor.

About 80 percent of situation comedies are sweetened by one man in Hollywood, Charles Douglass. Or rather two men, for Douglass has shared his closely guarded trade with his son. Exactly how the

Douglasses achieve the wide variety of laughs necessary to sweeten a show is unknown, although it does involve the use of a large black box which Douglass devised in the 1950s while an audio engineer at CBS. The box is operated with a keyboard for the different types of laughter, a knob for volume control, and a foot pedal. Inside the box presumably are tape loops of chuckles, snorts, titters, laughs, and yowls which the Douglasses have collected from studio audiences over the years. These sounds are folded into the sound track of the show they have been hired to sweeten. If there had been a studio audience, the sweetening is added to what is already there, but if the comedy had not been filmed before a live group, the Douglasses will dexterously create the laugh track from scratch.

Russell Baker, the droll *New York Times* columnist, got thinking about the laughs recorded years ago which still are heard on television shows. Tongue-in-cheek he wrote, "Many of the hardest working dead entertainers are those unsung and faceless laughers who toil nightly in television situation comedy. These are the people whose laughter you hear after the boffolas on shows that have been filmed without audiences. I don't suppose all these laughers are dead, but a lot of them must be by this time.

"I know for a fact that at least one is, because he is my Uncle Parker, who died in 1941. He was famous in the family for his distinctive laugh. During the 1930s he traveled to New York and attended a radio show. Eddie Cantor, Joe Penner, or somebody like that—I don't remember—I was still very young.

"As Aunt Emma still recalls, 'Parker laughed himself to death.' A recording of that show with Uncle Parker's distinctive laugh must have been put in a studio attic and forgotten for years until the birth of television gave birth to the situation comedy, which gave birth to the so-called laugh track, which put poor dead Uncle Parker into millions and millions of American homes."

As well as adding laughs, sometimes sound engineers like the Douglasses subtract studio laughter, if it is inappropriate or too prolonged to be believed by viewers. It wouldn't do, for instance, to have a gleeful display when Lucille Ball first strode onto the set of a new episode. The studio audience appreciated the actress's appearance, but the home audience would be looking for an unapplauded character. Much of the art of sweetening consists of toning down and evening out the responses of a studio audience.

Why is a sweetened laugh track needed? Producer Sam Denoff of *On Our Own* said, "Laughter is social. It's easier to laugh when you're with people. In a movie theater, you don't need a laugh track, but at home, watching TV, you're probably alone or with just a few others. A laugh track creates the atmosphere of an audience. It helps you respond." Scientific research supports his view. If canned laughter is added to a show, a test audience will laugh along more frequently and for longer periods of time than if the canned laughter is absent. Under experimental conditions, when situation comedies are graded on how funny they are felt to be, higher marks go to the version which incorporates fake laughs. Sweetening unquestionably helps to bring forth the release of laughter.

It can be different for Americans to concede that laughter is a purging of hostile impulses, and that their favorite form of televised fantasy is not as innocuous as it first appears but is pinned on ten or twelve little invitations to aggress. In this country people choose to find nothing venomous about situation comedy. It's the particular cultural filter which Americans bring to the subject of comedy that permits seeing it as untainted. Whole books are written by Americans about television's influence without any mention of situation comedy, so free is that brand of fantasy from the reconsiderations and controversies that dog other types of television content.

The Japanese perceive comedy differently. They don't have the cultural filter which allows them to ignore the viciousness inherent in joking. Sensitive to the demeaning aspects of humor, to the laughingstock's loss of standing, the Japanese don't like to see comedy on their television sets. Although the video habit has pervaded Japanese life even more than American, situation comedy has not.

But for Americans, cheerfully disregarding the violence at the base of humor, situation comedy is something to be fully enjoyed for the release and pleasure it affords.

The Mary Tyler Moore Show

In the 1970s it was rare for a situation comedy to be so widely popular and yet so innocent in tone as *The Mary Tyler Moore Show*. On Saturday evenings when Americans ought to be kicking up their heels, the serial brought into millions upon millions of households a half hour of comedy which could hardly have been more inoffensive. Yet if we scratch below the surface, we'll find that this situation comedy too dealt in aggression.

The beguiling virtuousness of the show was an aura created by the central character, Mary Richards. Whether at the now legendary news-room of WJM or back in her apartment building, Mary brought care and goodwill to seven seasons worth of weekly mix-ups. Whoever came into contact with the associate news producer was bound to be received with grace if not deference. Viewers expected Mary's behavior would always be above reproach; the very few times it was not, there was sure to be a good reason.

CBS vice-president Perry Lafferty offered this view of why Mary Richards struck such a chord with Americans: "She's the well-scrubbed, all-American girl that everyone likes. I think it's her vulnerability that makes her particularly appealing. Little girl lost. Also, she's beautiful and all that without being threatening." Her vulnerable, sensitive personality was revealed in every gesture and expression. While Lucy's comedy was broad, even garish, Mary's was restrained and finely done. Her face disclosed not a schemer's delight but oftentimes a soul in conflict, as opposing feelings rose at the same instant. Lucy initiated conflicts, but Mary was sought out by them—Ted Baxter had a heart attack on the air, Murray Slaughter missed a newswriting award, Lou Grant got a divorce. In that she was more acted upon than she acted out, many viewers may have seen something of their own lives.

But whatever happened during the half hour, in the end Mary was not to be unhinged by it. She survived, she endured, and occasionally she triumphed. Viewers spotted in her the grit they hoped to discover in themselves. More often than not it was due to Mary herself that things were set to rights and the order of this situation comedy's universe was restored. Her essential attribute, the means by which she confronted the world, was her penchant for order. "I always do my hair before I go to the hairdresser," she once remarked.

Americans found Mary Richards engaging because the character was so ideally played by the actress Mary Tyler Moore. There was a close identity between the two which extended far beyond first names. Miss Moore described herself one time, "There are certain things about me that I will never tell to anyone because I am a very private person. But basically what you see is who I am. I'm independent, I do like to be liked, I do look for the good side of life and people, I'm positive, I'm disciplined. I like my life in order, and I'm as neat as a pin." Her campaign against disarray was testified to by Valerie Harper (who played Rhoda) when talking about the difficulties of producing *The Mary Tyler Moore Show* week after week: "Mary's presence creates order."

In the way that *I Love Lucy* was thematically suitable for the audience of the 1950s, *The Mary Tyler Moore Show* was appropriate for the 1970s. *Lucy* had offered a fantasy setting where the tensions of America's rekindled family life could work themselves out, where the war of the spouses could take place and nobody would get hurt. Twenty years later the battle lines had shifted from wife versus husband to person versus world, and Mary Richards had become the standard-bearer.

Demographically, viewers in the '70s were younger and less settled into their adulthoods. Marriage was not the bolstered institution it had been—many were delaying their entrance into it, and those who were installed were often restless, to the point of getting out. Divorce statistics zoomed as marriage rates decreased, and even the traditionally high remarriage rates began to falter. The baby-boom products were growing up at a slower pace than people ever had, as if their sheer numbers were enough to reinforce and prolong the state of adolescence. More given to mating than marriage, the audience was not adverse to a situation comedy starring someone who was single.

The dilemmas of the early '70s, in brief, were not so much those within the family as those outside. Questions of finding one's way, particularly through the world of work, occupied the minds of many. The challenges and pitfalls of employment, the stresses of life in large organizations, confronted Americans by the millions. For those who were getting their toes wet in adult careers, as well as for others younger and older who empathized with the energetic but unsure young adults, tensions were being created which demanded outlets.

If a cold shoulder was being turned toward marriage, and a sweaty hand was being offered toward employment, all of this was going on against a backdrop of general uncertainty and potential chaos. The New Left was continuing to pose a threat to the order of things, but the excesses of the Nixon reaction were even more troublesome. A recession was doing its damage, revealing, along with the oil embargo, how infirm the underpinnings of modern existence might be. Turmoil was never outside the realm of possibility.

Order was what *The Mary Tyler Moore Show* proferred. For those who were unmarried, and for those who were negotiating their way up the first rungs of their careers, the show had a great attraction. In the face of uncertainties, *Mary* offered up a tidy and sustaining fantasy world. Single life was not desolate, work was not oppressive, and the world was not teetering. "You're going to make it after all," the show's theme song promised at the start of each episode.

The show accompanied and complemented the changes going on in American life, but it did not precede them. Mary Tyler Moore told an interviewer about the show, "It has, apparently, made an awful lot of single women who had been ashamed of being alone and dateless on Saturday nights suddenly very happy with themselves, content to be alone." But by the same token, the actress explained on another occasion, it was not the business of the show to instigate change: "I was on a panel once, discussing 'women and television' with Gloria Steinem and several women's liberationists, and someone in the audience asked me why I didn't push for scripts with more to do about equal rights. Well, sure, I believe in them and we'll always sneak things in where we can, but the show isn't about furthering causes, and if it were, it wouldn't be funny. The show is about six characters and their relationships."

These six characters—Lou, Ted, and Murray at work, Rhoda and Phyllis back home, with Mary in the center—comprised the program's continuing situation. If ever a television station existed on cooperation and camaraderie, it was WJM. The sole reason the people in the news-room were there, it seemed, was to help each other out. And in the same fashion, at Mary's apartment building, Rhoda and Phyllis—as pungent as they were—would always be her friends and she would always be theirs.

So in terms of the relationship of the program to its audience, *The Mary Tyler Moore Show* was a most agreeable place to be. The underlying theme was pertinent to the lives of many individuals in the early '70s, and the central figure had a mix of insecurity and strength that permitted viewers to identify with her. On top of this, the repeating situation was a welcoming, supportive one. Conditions were right for viewers who entered this fantasy to let down their guard a little and laugh.

To see how the program drew out laughter, let's turn to a sample episode. This particular segment was directed by the man who directed most of the shows over the serial's seven-year run, Jay Sandrich. Years before Sandrich had begun his career as second assistant director on *I Love Lucy*, and there he had learned the filming techniques pioneered by Desi Arnaz. Sandrich's use of three cameras and a live audience to keep things frothy became ingredients in the freshness and success of *The Mary Tyler Moore Show*, just as they had been for *Lucy*.

Not far into this episode Mary and Murray are standing behind their side-by-side desks, talking with Ted. Mary has reluctantly asked Ted to speak to a ladies' club, and Ted has accepted with boyish enthusiasm.

Murray, who will have to write the speech, is less keen, saying, "What's Ted supposed to talk about?"

"Oh, what difference does that make? They'll love me," gushes Ted. "I've gotten cheers by just cutting a ribbon at a supermarket opening."

"That's because they didn't think you could do it," cracks Murray. Laughter from the audience is the next thing heard as Ted fidgets.

Humor on *The Mary Tyler Moore Show,* as on any situation comedy, depends on jokes directed at laughingstocks. The perpetual butt on this serial is the egomaniacal anchorman. It's difficult to imagine a character more conceited and more deserving of ridicule; the show's creators and writers have made sure of that. The audience can laugh without reservation as Ted gets what's coming to him. Aggression is being discharged guiltlessly.

In this particular episode enough hostility flows toward Ted that he is bowled over. It all begins when Phyllis convinces Mary to get Ted to speak to her club. The scheduled speaker cancelled out, Phyllis explains. Here the writers slip in a sexual joke, for the defecting speaker was not an authority on flowers or home decorating. "We had Dr. Herman Davis lined up, the controversial psychiatrist who wrote that great book, *Don't Be Embarrassed About S-e-x.* But he backed out. He decided at the last minute he was uncomfortable with women."

Ted does deliver the speech, but things do not go well afterward. What transpires we learn from Phyllis at Mary's much later that evening (a dramatic economy, since the show rarely used sets other than the WJM newsroom and Mary's apartment). The question-and-answer session following the speech (first question: "Are you for or against women's liberation?") has proved to be too much for Ted, and he has been reduced to a blithering idiot. The recounting of his downfall elicits more and more laughter from the studio audience. When Ted mentions that his mother and father had been in the audience and that, "They were the first to leave," the audience howls.

Not all the hostility on *The Mary Tyler Moore Show* is directed at Ted. Some of the humor occurs at the expense of Lou Grant as the *paterfamilias.* And at times Mary catches a bit of it. In the aftermath of Ted's collapse, Lou tells Mary, "I'm one of the few people in my field who doesn't have a peptic ulcer. And one of the reasons for that is, I am able to delegate blame. Nothing that goes wrong here is my fault. It's Ted's fault, it's Murray's fault. *This* is your fault. *So fix it, before it looks like my fault.*" And the laugh track crackles.

Rebuilding Ted's confidence becomes the dramatic issue of the program. A butt that goes limp can't be satisfying for long. It's like a Bobo doll that loses its air and refuses to roll back upright for the next blow. Only if Ted were back on his feet would the fundamental order of WJM be restored and the following week's humorous episode be possible.

The solution is to introduce yet another laughingstock into the proceedings, for Ted's ridiculous problem is best solved by a similarly ludicrous figure. In wanders a hang-dog publicity man, Dave Kierson—the only PR person ever to have his phone number omitted from his business card. Trying to publicize the Midwestern Yo-yo Manufacturers' Association, Kierson's idea is to give a Man-of-the-Year award to a newscaster in return for mention over the airwaves of his client. Mary, eyes lighting, nominates Ted. When Ted receives the award, his confidence returns with it. As the episode ends, he is demanding that Lou have his dressing room repainted. The world of the WJM newsroom has been tidied up once more.

And the audience feels relieved. Trapped impulses have been pried out of viewers by the jokes, and the amount of aggressive feeling stored in minds has been reduced. People would already be looking forward to the sense of warmth and relaxation the following week's story would bring.

In television's armamentarium, situation comedies like *I Love Lucy* and *The Mary Tyler Moore Show* are the best medicine for psychic purging. They usher viewers into a padded fantasy world where laughter can come easily. For the greater number of Americans, situation comedy draws out the more readily dischargeable tension. It can't get it all out, however. Another kind of video fantasy is needed for the deep scouring.

7
Television Is Good for Your Spleen

Violence Studies

Fifteen-year-old Ronnie Zamora of Miami, Florida, was a devotee of violent television programs like *Kojak*; he also shot his elderly next-door neighbor to death. The connection between these two facts remains tenuous, debatable, and of the utmost importance. If television shows can be proven to broadly set off violent behavior, then we would have the elusive key to many of our most obstinate social problems.

As seen, Ronnie's lawyer asserted there was a cause-and-effect relationship, that exposure to television violence had precipitated a murder. But the contrary point of view, that video brutality had little to do with real-world crime, seemed more reasonable to the jury, and they convicted Ronnie. Whether they cared to or not, the jurors had taken a stand in the prolonged debate on the effects of televised aggression.

It's a debate which alternately passes through periods of great fervor and then half-decade lulls. Against a backdrop of social upheaval and general rambunctiousness, the issue was heatedly discussed in the early '70s. When the Report to the Surgeon General on Television and Social Behavior came out in 1972, many thought it amounted to more ammunition for Media Snobbery's assault on television violence. Even though a close reading of the Report reveals it to be less than decisive, in the heat of the battle the several condemnatory studies were brandished about. But by the end of the '70s the disagreement was less feverish and strident. A general impression prevailed that violence on the medium had subsided, and this took the edge off the critical attack. Certainly, all concurred, the number of action/adventure shows had declined.

George Gerbner, who carefully counts up television's violent incidents for his annual "Violence Profile," did not agree that media mayhem was skidding as the '80s began. The gist of his reports was that there was just as much violence as there had always been, but that it was more evenly distributed through the prime-time hours, and better camouflaged. While he didn't say so, it's possible that what may have been lost in the shrinkage of action/adventure programming was made up in the expansion of force-filled sports broadcasting and made-for-television movies. In short, what seemed to be varying was not so much the amount of television violence as the heat of the debate surrounding it. Since aggression appears to be an intractable component of television, some experts found it easy to foresee the resurgence of the action/adventure format.

Whether the debate flares or smolders, the question remains the same: what is the relationship between televised aggression and real-life brutality? There are several possible answers: fantasy violence instigates real violence; or fantasy violence has no influence on real violence; or, conceivably, fantasy violence could reduce real violence through the cathartic expulsion of viewers' hostile feelings. When an important question like this finds so many responses, it's only logical that people would turn to science for a definitive answer.

A succession of communications scientists dating from the first days of television have executed studies into the matter of media aggression, and the larger proportion of them have thought they found evidence of fantasy violence stirring up interpersonal violence. Albert Bandura, the researcher who had toddlers swat Bobo dolls, was followed by tens upon tens of other scientists whose methodologies have varied but whose conclusions have been largely the same. Since these media scholars, like most people, operated in the belief that social violence was on the rise and that television might be the cause, their work took on an air of grand purpose and high moral tone. So resolute were they that many members of the public now accept it as truth that assaults on the television screen translate into assaults in real life. As Russell Baker wrote in his column on the Ronnie Zamora trial, "Indeed, the belief that television violence breeds social violence is so widely held that to question it seems eccentric."

Representative of this string of studies is one conducted for the Report to the Surgeon General by Jennie McIntyre, James Teevan, and Timothy Hartnagel of the University of Maryland. The three researchers decided to forego laboratory experimentation (like Bandura's work) since such efforts were increasingly the object of criticism. A laboratory is an unnatural setting for children, and the young subjects may have behaved unnaturally as a result. Findings might not have described how they would actually behave in the real world. If the children did become aggressive after seeing aggression on the screen, it could be because the tots thought that was expected of them from the hovering laboratory personnel.

In an uncomplicated study the researchers turned to real-world data. Over 2,000 junior and senior high school students, both male and female, were asked to list their four favorite television programs. The program choices were interpreted in the light of violence ratings which the research team had previously established for all broadcast shows. Each of the 2,000 respondents was also asked to complete a checklist on his own deviant behavior, which contained questions about violent acts, petty delinquent behavior, defiance of parents, political agitation, and involvement with the police. Then the viewing habits and the self-reported deviant behavior were compared. The conclusion was that "there is consistently a significant relationship between the violence rating of four favorite programs and the five measures of deviance." The students who best liked action/adventure shows were the students most inclined to be violent and delinquent. The implication was one of cause and effect.

Before digging into studies like this one, and penetrating beyond them to the actual relationship between televised violence and real-world violence, it is wise to be familiar with the trends of violence in American life. The measure of social violence which comes most readily to hand is crime statistics. It is the Federal Bureau of Investigation that is the fount of these figures, and the annual message of their Uniform Crime Reports appears to be that the lid is just barely being held on.

Of the 29 varieties of crime reported to the FBI from local police departments across the nation, the seven summed for the well-known Crime Index are murder, forcible rape, robbery, aggravated assault, burglary, larceny of $50 or more, and auto theft. The Index, which is the total rate of these offenses per 100,000 people, has climbed since first issued in the 1930s, giving the impression that crime itself has been proliferating. The rising figures tend to make the FBI look ever more indispensable and worthy of funding. According to Fred P. Graham, a legal correspondent for the *New York Times* and contributor to the 1969 Report to the National Commission of the Causes and Prevention of Violence, "The FBI's stewardship of the nation's crime statistics has resulted in a hysteria that seems more beneficial to the FBI as a crime-fighting public agency than to the public's enlightenment."

The vagaries of data-reporting are one reason not to rely unduly on the FBI version of crime in America. Police department figures do not so much reflect the depth of criminal behavior in society as other, more prosaic factors like available manpower, the education and training of those who collect the numbers, local interpretation of what constitutes a violation, municipal politics, and so forth. The FBI is not at fault for this, but it is a situation which invites them to make use of it for their own purposes. Local departments have come to sense that it is higher numbers which most interest the Bureau. New York's robbery figures used to be several times lower than Chicago's until the FBI stopped accepting the east coast city's reports; when the numbers shot up, New York was reinstated.

Inflation is even more of a reason to be skeptical about the FBI's picture of crime trends. More than 80 percent of Index crimes are thefts of property, and among them larceny (counted as $50 worth or more) is most likely to be influenced by the rising costs of goods. To steal something valued at $50 was a considerable theft in the 1930s, but is almost an inconsequential one today. As dollars decline in purchasing power, the number of cases that fall within the limit for larceny can only increase. Two hubcaps stolen three decades ago would not be counted; nowadays they would be.

Certain social changes also call the Crime Index into question. Rape went largely unreported in previous years because it is so degrading to the victim. But in an era of new, less cowering attitudes, a higher percentage of violations is reported.

The major social change affecting crime statistics has been the integration of minorities into the mainstream of American life. When blacks own more insurance and make more use of income tax deductions, more thefts are reported. In addition, along with other social services, police protection, too, is being extended into ghettos, and, as a consequence, more arrests are made. This does not mean more crime is taking place there; in fact, it may well signal just the opposite.

The FBI figures that have struck fear in the hearts of Americans may have done so needlessly. Long-term trends in crime, going back a century, are the opposite of what the Crime Index implies. While there are no figures for the entire nation in the 1800s, good municipal statistics have been collected. Social scientist Theodore N. Ferdinand has computed the crime trends in Boston from the middle of the 19th century to the middle of the 20th by culling data from the annual arrest reports of the Boston police departments: "The aggregate crime rate in Boston has shown an almost uninterrupted decline from 1875–78 to the present."

Another way to appraise the true trends in crime is to abandon aggregated rates with their inflatable components, and look at just one highly calculable crime. Murdered bodies are a very good unit of measure—they are only infrequently concealed with success; they are eminently countable; because they differ from money and do not inflate or deflate, their numbers lend themselves to comparisons over time. Ferdinand reports that the Boston homicide rate, which was over seven deaths per 100,000 population in 1855, fell by 75 percent to a rate below two per 100,000 in the middle of the 20th century. The federal government's U.S. National Center for Health Statistics has kept tabs on country-wide homicide rates during the 20th century, and states that a peak was reached in 1933 at 9.7 per 100,000, thereafter dropping to 4.7 in 1960.

But by any measure, something strange happened in the 1960s to America's long-term descent in crime and violence. The figures turned upward again. Crime in all categories rose slowly until 1967, and then more swiftly. Homicides climbed to 8.3 per 100,000 by 1970. Suddenly the FBI figures began to look not so puffed out of proportion. What was going on to produce this abnormal rash of crime? Could television somehow be behind it?

The answer turns out to be simpler than Media Snobs might care to admit. Statistically, crime is the handiwork of adolescents. Compared to other age categories, teenagers get arrested at rates five to ten times higher. Seen this way, a very significant fact about Ronnie Zamora was the curt but informative one of his age. What a demographer would see as the greatest change going on at the end of the '60s was the arrival into adolescence of out-sized groups of baby-boom children. The children produced during the period of revived family life were passing on up through the age levels, and were coming en masse upon their 14th, 15th, and 16th birthdays. The sheer numbers of them were enough to cause the crime rates to soar.

Criminologists and demographers taking stock of this state of affairs felt safe in predicting that crime figures would stop ascending once these cohorts had passed into their twenties. This is what has occurred. Homicide rates peaked in 1974 and turned down the following year. By 1977 the FBI admitted that crime was slackening: the Uniform Crime Report for that year told of a 4 percent drop from 1976. The rates continued to fluctuate through the rest of the '70s as the remainder of the baby boom exited adolescence, while officials who draw up long-range law enforcement budgets or plan for new penitentiaries acknowledged that a lengthy descent was in the cards. "By the time new prisons are planned, designed, and constructed, much of the need for the extra capacity is likely to have disappeared," wrote urban planner Alfred Blumstein in 1980.

It had been the odd post-World War II demographic profile of the United States, and not television, which had put a bump in the century-old slide in crime rates. But what about a study like the one by McIntyre, Teevan, and Hartnagel which had found a correlation between violence-viewing by teenagers and self-reported delinquent behavior? Doesn't that indicate what's to blame? It might if the conclusion were true. Three years later the researchers reanalyzed their data, however, and were compelled to do an about-face. The correlation they had found the first time out was not there after all: "We are forced to conclude that the TV violence predictors, both objective and perceived, do not matter significantly in explaining violent behavior." They confessed bravely, "Our results suggest that TV violence does not significantly affect actual violent behavior."

(In passing, one of the things this research team noted was that, no matter how much violence there is on a favorite show, it is rarely perceived by the fan. People aren't able to recall the brutality on the shows that absorb them. This fact will prove to be informative about the true function of action/adventure programs.)

Their belatedly revised finding is in agreement with a set of studies done back at the same time the three researchers were working on their initial, misguided analysis. Not contributors to the Report to the Surgeon General, City University of New York psychology professor Stanley Milgram and his associate Lance Shotland were also interested in the relationship between media violence and real-world viciousness. As described in their book *Television and Antisocial Behavior,* they executed a lengthy series of experiments using hundreds of subjects. In order to study the effects of television under natural but controlled conditions, they arranged to have made several special episodes of the series *Medical Center,* identical except for a scene or two. The basic plot concerns a medical attendant at the Center, Tom Desmont, who quits his job but then, as his financial situation worsens, asks to be rehired. His boss, Dr. Gannon, tells Tom the position has been filled. Subsequently, Tom sees Dr. Gannon on a telethon informing viewers that collection boxes for a new community clinic have been placed around the city. As one version of the specially filmed drama proceeds, the desperate Tom is driven to smash open some of the collection boxes.

Milgram and Shotland assembled audiences for these experimental programs in a variety of ways, including handing out invitations on New York City street corners. Promised they would receive a free transistor radio for attending, people were shown an episode in a midtown theater used for previews. The viewers were then instructed to go the following day to a nearby office where they could pick up their gift. When they arrived there (their arrivals were staggered, so that only one person entered the office at a time), they found an empty room with a sign saying there were no more radios to be distributed. Also in the unattended office was a collection box for Project Hope, similar in design to the one in the television show, with money clearly visible inside it. The behavior of the frustrated subjects was surreptitiously videotaped. On the way out signs directed them to another office where they were in fact given their free radio.

Were people inspired to imitate breaking into the charity bank, as they had seen on the episode of *Medical Center?* Those who saw the segment in which Tom smashed the boxes were no more likely to go after the money than those who saw the segment with that scene omitted. Trying to improve on procedural matters in the light of what they learned after each experiment, the two scientists conducted a total of eight studies, but each time it came out the same way, with depicted violence having no influence. "We did our best to find imitative results, but all told, our research yielded negative results," they concluded their report.

What was discovered in individual probes like this effort by Milgram and Shotland, or McIntyre, Teevan, and Hartnagel's revision, is also the conclusion of several subsequent overviews of the entire multitude of violence studies. R.M. Kaplan and R.D. Singer, two psychology professors at San Diego State University in California, surveyed the violence studies done before 1976, sorting them into three categories: the *activation* view that violent fantasy content causes aggression, the *catharsis* view that aggression would be decreased, and the *null* view. It was the null position which was best supported by the literature, they reported— television violence is unrelated to social violence. In any case, "It is our argument here that the evidence that TV causes aggression is not strong enough to justify restriction in programming."

Two British scientists, Dennis Howitt and Guy Cumberbatch, poured through the same body of articles and books as had Kaplan and Singer, and added to it the many studies on media violence done on their side of the Atlantic. When all was said and done they wrote in their book, *Mass Media Violence and Society,* television could not be convicted. "The central theme of this book is that the mass media do not have any significant effect on the level of violence in society."

The pendulum on television and violence studies has swung back toward the midpoint, to the view that there is no connection between media aggression and social aggression. Given the prevalence of Media Snobbery among academics, it would not be surprising to learn that no bold scientist was going even further and exploring the notion that televised violence could actually relieve aggression in viewers through catharsis. But there is someone, and he has been pursuing this line of investigation in a careful and scholarly way since the 1950s. His name is Seymour Feshbach, and currently he is the Chairman of the Psychology Department at the University of California at Los Angeles.

Seymour Feshbach

For Seymour Feshbach, the results of the most extensive study he had ever conducted were clear-cut. They revealed "that witnessing aggressive TV programs reduces rather than stimulates the acting-out of aggressive tendencies in certain types of boys." As related in his report *Television and Aggression,* among youths inclined to delinquency, a diet of violent action/adventure programs actually decreased hostile behavior. This finding was firm verification of the hotly debated catharsis theory— "catharsis" from the ancient Greek word for cleansing or purging, implying that viewed violent fantasy may serve almost as well as actual violence in ridding people of hostile impulses. Feshbach did not come to this conclusion easily, since for him as for most scholars it ran counter to conventional wisdom on the subject. The catharsis interpretation of television violence arose unbidden from a series of studies he had carried out over several decades.

In 1955 Feshbach published an experiment based on work done for his doctoral dissertation in psychology at Yale University four years earlier. The article had nothing to do with television; Feshbach's major area of research, then and now, is in the relationship of aggression and fantasy. This first experiment was designed to get at a very rudimentary question: does fantasizing relieve hostility? Freud had said so, but experimental results were lacking. To properly test the theory Feshbach divided 250 college students into three groups. The first group was to be roundly insulted before being given the opportunity to indulge in fantasy; then their hostility would be measured. The second group, also insulted, would have no chance to fantasize. To control for extraneous factors, a third group would be exposed to neither insults nor fantasy before taking the test for hostility levels.

To stimulate hostility, Feshbach had a brusque colleague play the role of an irritating, outside experimenter. Upon being introduced to a class the sham experimenter would begin to berate the students (in words more pertinent to the '50s than later years): "I realize that you students or should I say you grinds, have few academic interests outside of your concern for grades. If you will try to look beyond your limited horizons, your cooperation will be useful. In other words, I'd like you to act like adults, rather than adolescents." He kept it up until the students were good and mad. Then the fantasy group was encouraged to make believe by creating tales based on ambiguous drawings shown to them. Aggression levels in all groups were tested at the end by a sentence-completion test and a questionnaire regarding the insulting experimenter.

The students who had the chance to fantasize were decidedly lower in the aggression ratings than those in the other group who had kept their resentments bottled up. The fantasizers had daydreamed their stresses away. If this were the case where fantasies were generated within the mind, Feshbach went on to ask, then would it also be true if the fantasies were supplied from outside? That is, if the fantasies were those of popular art? He set out to test this.

Published in 1961, the second experiment resembled the first in that the subjects were male college students, half of whom were subjected to the insults of a bogus experimenter. There were four groups this time: the first group was insulted and then shown a segment of fantasy aggression, a prize fight sequence from a movie; the second group was also insulted but saw a non-fantasy industrial film on the spread of rumors in a factory. The third group was *not* insulted and saw the aggressive film. The fourth group was likewise not insulted and was shown the neutral film. Aggressive levels were gauged at the end by a word association test and, as in the earlier experiment, a questionnaire about the insulting experimenter (who had long since left the room). The findings confirmed the hypothesis that the insulted group who saw the aggressive film footage were less hostile afterward than the insulted students who had viewed the neutral documentary. A violent filmed fantasy had reduced violent sentiments.

These two experiments were prelude to the major study which Feshbach, together with his colleague Robert Singer, executed in 1969 and published in 1971 as *Television and Aggression: An Experimental Field Study*. Network television shows were the fantasies being investigated this time. Would they accomplish what the aggressive movie fantasy had done in the 1961 study, or the self-generated fantasies of the 1955 experiment? Or would they, as most scholars insisted, only arouse aggressive feelings?

To see what the effect of violent television fantasies might be on viewers, Feshbach and Singer used male subjects younger than those in the earlier experiments, feeling that they would be the most sensitive to media aggression. If positive or negative effects were to be found, they would be most pronounced among this age group. The total number of subjects was greater than before—nearly 400 boys in all.

A primary difference between this study and the two previous ones was that this time out Feshbach's method was not that of a laboratory experiment but rather, as the subtitle of the book notes, a field study. Along with others studying the question of television and violence, Feshbach had become aware of the crippling liabilities in the methodologies of laboratory experiments, his own and others. Such experiments, being completely under the control of the experimenter, had a way of conforming private biases. Also, there was always the plaguing question of how much laboratory findings could be generalized to the world-at-large, where television viewing and its effects occur within a full-blown social context. Naturalistic field studies looked sounder by comparison. However, in a true field study it would be difficult to control the television diets of the boys or to observe their subsequent behavior. Feshbach and Singer hit on an ideal solution for their purposes when they decided to use boys' boarding schools. Here was a real-world situation which nevertheless permitted close overseeing of what the boys watched and how they behaved afterward. A total of seven institutions were recruited—three private schools and two homes for wayward boys in the Los Angeles area, and two more boys' homes near New York City.

Each of the 400 boys was randomly assigned to either a television menu high in violence or to a second one featuring nonaggressive shows. Boys assigned to the aggressive menu could choose from *Colt 45, FBI, Have Gun, Will Travel, Rifleman, The Untouchables,* and so on. The shows on the other list included *Andy Williams, Camp Runamuck, Gilligan's Island, Hazel, Lassie, Leave It to Beaver,* and *My Mother, the Car.* Over the six-week duration of the experiment, a boy could watch as much television as he wanted, or was allowed to, as long as it came from his designated list. The only show on both lists was *Batman;* this occurred because some of the boys on the nonviolent menu threatened to strike if they couldn't see the then-popular cowled crusader.

Before, during, and for the week immediately after the six-week viewing period, the aggressive behavior of each boy was carefully watched. The house parent or teacher or adult most responsible for a boy would judge each aggressive act on a scale of one to four—one if provoked and mild, four if unprovoked and strong. When the scores were all in and tabulated, it was learned that there was little difference in aggressive behavior among the private school boys between those who had been exposed to the violent programs and those who had been restricted to the blander ones. But it was quite another story for the poorer, delinquent youths in the boys' homes. There the boys who had been watching the action/adventure shows were far less rowdy than their friends who had been on the nonviolent diet. Fantasy violence had quieted them down. In Feshbach and Singer's words, "Two major conclusions are indicated by the experimental findings: First, exposure to aggressive content in television does not lead to an increase in aggressive behavior. Second, exposure to aggressive content in television seems to reduce or control the expression of aggression in aggressive boys from relatively low socioeconomic backgrounds."

Why there should be a difference between the findings for the private schools and those for the boys' homes was one of the unanswered questions that led me to interview Professor Feshbach at his office in UCLA's Psychology Department.

What was the reason, I began, *that violent shows decreased the aggression of those in the boys' homes, but left the private school boys unchanged?*

"With the private school boys there was no difference; that's right," he responded. "Those on the violent television diet were no less aggressive than those on the nonviolent diet. They didn't become *more* aggressive either, I should point out. But this lack of effect seems reasonable to me. Television fantasy violence is going to be best at reducing aggression for those people who are most in need of it—in this case, the semidelinquent boys from poorer families who were in the boys' homes. Because of their lives, they had more aggression to discharge.

"There's another aspect to this," Feshbach continued. "One of the more interesting findings of the *Television and Aggression* study was that boys who pretested lowest in aggressive fantasies were those who were most relieved by watching violent television shows. And these subjects tended to be in boys' homes more than in private schools. They were the ones who in general did less fantasizing and more acting out. For people who are disposed to act aggressively, and who have minimal fantasy skills, the data suggest that violent television can be an alternative fantasy mechanism. It helps them manage their hostility, measurably so."

Much of your previous research has been into fantasizing, I said. *Is it the fantasy element of violent television shows that helps needy viewers reduce their aggression?*

"Definitely. Television fantasies supplement a person's own imagination, and help him discharge pent-up aggression in the same way that dreams and other products of the imagination can do."

Let me return to the study itself, and ask you how you happened to do it.

"Well, as you know, the area of research I've specialized in hasn't been television or viewer effects, but fantasy and its relation to aggression. But way back in 1960 I chanced upon a notice in the *American Psychologist* about a competition to design an experiment to do with television and youngsters. It was a Sunday evening, I recall, and the deadline was the next day. For some reason I thought, why not? So I stayed up that night and figured out the research design, and wrote it out in longhand. The next day my secretary typed it, and I called the people in New York who were sponsoring the contest and told them I would put it in the mail that afternoon. They said that would be all right.

"The first prize was to be $1,000, and the other prizes were $250. I didn't get the first prize, and to tell you the truth, I was disappointed. The research design was published in a collection of pieces about children and learning, and I forgot about it.

"I moved around a bit—from the University of Pennsylvania where I was then teaching, to Berkeley, Stanford, the University of Colorado, and finally in the mid-sixties to UCLA. About that time I got a call from the people who had sponsored the original design competition, asking me if I would be interested in actually carrying out the study. Once again I thought, why not? It took them several more years to arrange for the funding, and then it took me a couple of years to set it up. So almost ten years elapsed between doing the design and doing the study."

Was it a difficult study to set up?

"Yes, it was. It took a lot of time to search out institutions and find ones that would agree to let us invade their lives in a very significant way. We brought television into one private school that had never had it before. I'm still a little uneasy about that."

Once the study was done, what was the reaction to it among your colleagues?

"Some people who knew the book was coming out didn't want me to publish it. I got a phone call from one colleague asking me to halt publication. These are controversial matters, and I can understand people's feelings. Obviously it's hard to reconcile my findings with other data.

"After the book was out, I think on the whole it was well received and fairly reviewed. There was certainly strong critical scrutiny of it. When you get findings in social science that are inconsistent with prevailing expectations, people will examine your work much more closely. That's only reasonable. The one review I'm particularly unhappy about was included in the 1972 Report to the Surgeon General. I regard that review as highly improper. It misrepresented the study."

That's not too surprising, I said. *Your book came out in 1971, demonstrating that television can relieve hostility, and the Report to the Surgeon General was released a year later, tending toward the opposite conclusion, that television causes aggression. There must have been some tension between you and them.*

"No, not overtly. I've met most of the contributors to the Report at one time or another, and I find them to be pleasant people, as I hope they find me. Anyway, I contributed a small study too, so I couldn't really condemn the whole group."

What accounts for the difference between your findings and theirs? In fact, between your findings and those of many researchers into television and violence?

"One answer, of course, is that my study was simply a fluke. I don't believe that's the case, though. It was too substantial, and too carefully done and reviewed, to be in error, in my judgment. A better answer is that most research demonstrating that television can stimulate aggression has been done in laboratories rather than in the field. Experimenters working in laboratories are more likely—unconsciously, I think—to structure things so that their personal passions and convictions will be substantiated. Just the fact that experimenters in this area have often chosen a one-minute film rather than a longer, complete dramatic episode is an indication of this. The one-minute filmstrip of pure violence is most likely to provoke aggression. There's no possibility of working through or managing the aggressive impulses that are elicited. A longer episode, like a full television show, has a beginning, middle, and end. The viewer can pass through the conflict to the resolution, and experience relief instead of frustration.

"With longer episodes, the laboratory experimenter can still determine the outcome of his experiment. He does this by the particular film he selects. Let's say one experimenter chooses the movie *Body and Soul.* John Garfield is supposed to be throwing the fight, but in the end he knocks down his opponent and retains the championship. The movie ends on a high note—his girl friend rushes to his side, the music rises. Subjects looking at this will probably feel good. On the other hand, let's say another laboratory experimenter chooses *The Champion,* as several have. Kirk Douglas plays a very unpleasant character who gets beaten up in the end. Subjects watching this film are left with unpleasant feelings. So here we have two different stimuli producing two different responses, although objectively the amount of violence is the same.

"What I'm saying is that it's easy to make a laboratory experiment conform to one's beliefs, and that it's much harder to do this with a field study. A field study such as this one can come up with some interesting findings."

Would you agree that behavioral science is shifting away from the position that television causes aggression?

"I think it's safe to say that. A number of studies have been coming along with no findings one way or the other. I think there is a movement in the direction of more field studies, and toward the realization that television violence does not stimulate aggression. I feel confident that in time we'll see more and more studies which find a reducing effect."

What are the policy implications of your study?

"I want to be very careful here. I wouldn't want anybody to draw strong social inferences from one study, particularly when there were so many other studies reaching the opposite conclusion. I'm not going to propose that criminals should be compelled to watch televised fantasy aggression. I would say simply that action/adventure programs are hardly likely to be sources of stimulation of aggression. I think the case for censorship is very weak."

What do you make out of the efforts by groups like the Parent-Teachers Association and the American Medical Association to reduce the amount of violence on television?

"On the part of the AMA, I think it's totally irresponsible. They are venturing beyond the areas where they have data, and are operating without a real scientific basis. I think that's poor policy. It detracts from areas where they do have data, such as smoking and heart disease and so forth. As for the PTA, I wonder what makes them join in at this late date, when they never voiced these concerns before. Anyway, I think it's a displaced effort on their part. If they really wanted to do something about violence, if they really wanted to reduce delinquency, they should work to keep school facilities open longer in the day. It's that simple."

Kojak

The camera focuses in upon an armored bank car in the financial district of Manhattan. Suddenly a gang of men wearing ski masks surround it; the surprised guard is thrown to the ground. But before they can make good on the robbery, a squad car screeches up. Gunfire is exchanged, and one of the robbers is shot. The thieves escape in their getaway car, pursued by more police cars. The race through the city streets ends when the escape car bangs into a mound of cartons and comes to a halt. In the gun fight that ensues, policemen are wounded. One officer is captured by the four remaining robbers and dragged into a nearby store. This is where the criminals will make their stand—a sporting goods store with enough firearms and ammunition for them to hold off the New York Police Department for a long while. The sound track is filled with the discharge of bullets. On the police radio the call goes out for Lt. Kojak.

This was the rapid-paced opening action on the first episode of *Kojak,* premiering October 24, 1973. It was the sort of violence which Ellis Rubin, attorney for Ronnie Zamora, said had undermined his client. Of all television action/adventure shows, *Kojak* was the most vicious, Rubin indicated. Others agreed: a poll taken in 1977 of 2,000 viewers 12 years old and up concurred that *Kojak* was the most violent program on the air. No other show inspired so much ire from the Parent Teachers Association; in its well-mounted campaign against the medium, *Kojak* ranked as "the worst program on national television" for its brutality. Even the very name "Kojak" was too much for the sensibilities of Media Snob Rose Goldsen, who referred to it as "the vocal equivalent of a snap of the fingers."

The lead character, Lt. Theo Kojak of the New York Police Department, was played by the shaven-head, extroverted Telly Savalas, the one actor that Ellis Rubin wanted to subpoena to testify at the Florida murder trial. Savalas once remarked about the role he created, "I like Kojak. I'd like him for a friend. I think you know where you stand with him, whoever you are. He's fair. He's going to look the other way once in a while. Men watch Kojak because they see a guy they'd like to be. And women see a guy who's strong but still wounded a bit, tough but not too tough, full of compassion." His toughness was the prime reason so many millions of men and women idolized Kojak. Men wanted to identify with a character who was certain, aggressive, potent—all the things that they couldn't quite manage to be in their daily lives. American women found the hairless, lollipop-sucking lieutenant sexy and enticing; they wanted to be in the presence of a man who dominated his world as Kojak did, and they were not shy about writing Savalas and telling him so. He was king of their reveries.

During the premier hour-long episode, Kojak is very much in charge. He is the one on the bullhorn to the crooks holed up in the sporting goods store, the one who goes inside to negotiate with them and to see to the condition of the wounded policeman being held hostage. The robbers give Kojak two hours to arrange for their escape by helicopter before they will start killing the hostages. Within this tension-engineered situation Kojak keeps a level head. The armored tank he had first called for to assault the hideout, he now restrains. Teams of policemen are delegated by him to dig through from below and break in from above.

Meanwhile, Kojak tries another approach. The robbery has been arranged by the Tolaba brothers, one of whom is barricaded in the store, while the other has gotten away. Kojak pursues the loose one, intending to bring him back to talk his brother out of his foolishness. The search for Frank Tolaba gives Kojak the opportunity to rough up some New York low life. But bringing the brother back does not solve the problem.

In the nick of time the police manage to break into the building. The officers cutting through the roof are repelled, but the ones tunneling into the basement manage to get through. Kojak, shooting and ducking from aisle to aisle, finally subdues the band. He reaches the dying officer whose last words are, "We're the only line between the people and the crooks." With the dead policeman in his arms, Kojak walks out of the store and past the respectful throng of reporters and bystanders.

For the audience, violent fantasies of this sort perform a much-needed psychological service. As with the youths in Seymour Feshbach's boys' homes, viewers' stores of hostilities and tensions have been reduced. They have accompanied someone they admire and can identify with through a harrowing adventure, much as if they had done it themselves. They have played at being shrewd and righteous, but above all they have acted aggressively. They have bellowed, slapped, socked, and shot. They have stormed the fortress of crime. Since it has all been done in the name of justice, no recriminations are forthcoming. Instead there are rewards to be had from the wonderment and praise in the eyes of the gathering crowd. Doing this in their imaginations, viewers have asserted themselves past the taboos and restraints of workaday life and have risen to the level of the heroic. They have set their spirits free for an instant by purging themselves of the burden of their animosities. The tribulations of life which can so forcefully bow us down have been thrust off.

It is clear why television fantasy-makers routinely resort to the character type of the crime-fighter and look for actors who project a certain brutality to energize the roles. Plumbers and claims adjusters aren't good models for heroes because it is not socially permissible for them to throw their weight around. But a crime-fighter, striking back against the inroads of evil and malice, is allowed to meet force with force. By identifying with the hero and indulging in sanctioned violence, viewers are vicariously able to lash out and release some of their own resentments at the impositions of their worlds. This they privately do by the tens of millions.

Sports

Hits, strikes, blows, tackles, and shots are not the currency of action/ adventure shows alone. Another category of television content, sports programming, also trades in various sorts of assaults. "The idea of conflict is central," observes television critic Horace Newcomb about sports. "Legitimate violence is present in varying degrees in all athletic contests." In those sports where people aren't hit, balls are the surrogates.

At first glance television sports may not seem like fantasy in the way action/adventure shows are. Most sporting events are televised live and are not, everyone hopes, scripted beforehand. They are apparently as real and as unpredictable as life itself. But sports, like fantasies, are the antithesis of the workaday world; and in the last analysis, both are play. The parallels between televised sports and televised action/adventure are several. Structurally, both are highly formatted within severe time constraints, and are thoroughly governed by conventions. Both tolerate a certain amount of novelty within the strong structures—the action must be more than boring but less than unbelievable. The contest in both cases involves good guys and bad guys asserting themselves against each other in the interest of victory. If our favorite good guys don't win often enough on the playing fields, it's not long before they are thought of as bad guys. Boos signal that.

Within the structure of the athletic contest, as within the structure of the action/adventure show, it can be a particular heroic player that makes all of it engrossing for the viewer. By identifying with the sports hero the viewer can hit, slug, plummet, and succeed just as the player does. Whatever Reggie Jackson or O.J. Simpson or Jimmy Conners accomplished, the viewer accomplished it along with him in the same way that, with another program at another time, he could aggress and triumph in the skin of Lt. Theo Kojak. The need to assert oneself and win has been obliged.

So strong is the desire of Americans to enter imaginatively into televised athletic contests that the three networks broadcast almost 1,200 hours annually of well-toned humans striking at things in accordance with the rules of the game. This amounts to nearly 400 separate events going out to a yearly total of five billion spectators—which means that on the average each man, woman, and child will view twenty contests per year. It doesn't actually happen that way, though; as with the action/adventure audience, a sports viewer is twice as likely to be an adult male rather than a woman or child. And although some may imagine it's a brutish, Neanderthal male who is the typical viewer, the truth is that the audience for sports is skewed toward the better educated and more affluent. This is why television advertisers are enamored of sports, for it is their chance to contact the males with money to spend.

The appeals of the advertisers to this audience are not received gladly. A 1978 poll commissioned by *TV Guide* found that what these relatively well-off viewers want less of on their televised games are commercials, and what they want more of is the chance for yet greater involvement in the action through more use of innovations like instant replay and slow-motion shots. They want to feel intimate with the contest and the leading players so that they too can experience the sensations of playing to win.

It has been charged that, in conveying the thrill of the contest as vividly as possible to an enormous audience, television has wrenched sports out of shape. Media Snobs contend that the televising of sports has unnaturally altered them in the interest of network profits. Games have been repositioned to suit network schedules—and have been restructured to allow for more commercial breaks. The response to this criticism has stressed that games are artificial contrivances anyway and that there is no reason they shouldn't evolve as the demands on them change. Considering the magnitude of the institutions being blended— professional sports and network broadcasting—the changes have been relatively minor, and often in the direction of greater viewer pleasure. Now great multitudes of people are able to enjoy what only a limited number could witness before the video age began.

Another charge made against televised sports is that they are rendering Americans more passive. Sitting home in the gloom and watching a very few fling themselves about is supposed to produce beer-bellies and an acquiescent temperament. In truth, the real influence of televised athletics appears to be in the other direction, toward inspiring more and more Americans to get out and do it themselves. The more sports broadcasting there is, the more athletic Americans have become. Golf and tennis are two games that most observers agree have hugely benefited from the contests seen on television. And one sport, gymnastics, would scarcely exist for Americans if television had not brought it into millions of homes.

But the most viewed sport is too vigorous to induce much imitation by its adult audience. For an American male's imagination, it is the brawn and tactics of football that is most engaging. In his mind's eye he can play the person who brings his team closer to victory—the senior, crafty quarterback, the crashing halfback, the darting pass receiver. Or, as sophistication about the game has grown, he might even identify with the defensive player who digs in and resists the opponent's encroachment. Either way, on the televised football field there are enough opportunities for aggressive fantasies to suit almost everyone.

Since 1970 the most watched regularly broadcast sports event has been ABC's *Monday Night Football.* There is something about that first day back at work each week that makes people want to see athletes bang away at each other that night. And want to listen to the most abrasive personality in sports broadcasting, Howard Cosell (who is, the *TV Guide* poll relates, at once both the best liked and least liked announcer in sports, topping both lists by wide margins). As Cosell says about himself, "There never has been one like me before, and there never will be one like me again."

The professional football season leads to television's contribution to the national holiday calendar—Super Bowl Sunday. Invariably the championship games will be viewed by the largest audience of the television season. What people will see, though, is more likely than not to be a lackluster contest played out by two overly cautious, exhausted teams. But its ritualistic importance couldn't be greater. It celebrates the culmination of the audience's half-year effort to fantasize their hostilities away. When it's over, we have only ourselves to turn to. *Time* magazine reported about the post-game relapse in 1978, "After a six-month diet of football, the American public must shake a national habit, and the transition is not easy. In the home of the Super Bowl Champion Dallas Cowboys, for example, police report more than twice the daily average of violent assaults on the Sunday after the football season ends. Spats between spouses can take a nasty turn. Old scores are apparently settled and, without the soothing football fix for fragile psyches, new grudges are formed."

Benefits

Kojak was not a favorite of American audiences alone. The Japanese adore the program, as they also relish *Baretta, Hawaii Five-O, Starsky and Hutch,* and all the most brutal of action/adventure series imported from the States. But while these shows are the most violent of American television, they are not especially gruesome standouts when compared to other Japanese programming. Not only are Japanese viewers offered home-grown detective series even more vicious than the imported ones, with stompings and beatings far outdoing Western imagination or execution, but their traditional dramas refashioned for the medium are drenched in blood and gore. "So-called costume dramas and old samurai-warrior tales invariably climax in an orgy of death," relates observer Charles N. Bernard.

These shows, which standardly feature such scenes as wildly violent rapes, are not seen by just a select few. As in the United States, television has permeated all echelons and regions of Japanese life. Media violence, then, is ingested in huge doses by one and all. And yet the Japanese people are among the most violence-free on the face of the planet. Crime has been on the decrease in Japan for 30 years. The homicide rate is miniscule and juvenile delinquency virtually does not exist. Asked to make sense out of this, George Gerbner, the creator of the Violence Profile, could only say lamely, "We are intrigued by the Japanese phenomenon, but we can't be sure whether all that violence is truly having *no* effect on the Japanese. It may be having a greater or different effect than is apparent." The obvious "greater or different effect" that Gerbner is unlikely to concede is that television fantasy violence is playing a role in the *reduction* of social violence.

Even more so than American life, Japanese society is a highly pressured one, severely governed by codes of etiquette and submission. Tensions and resentments could reach the breaking point if it were not for the therapy of nightly grapplings and hackings on television. By availing themselves of these fantasies much more openly and enthusiastically than Americans seem able to do, the Japanese divest themselves of some of their aggressions. It's a fact: violent television fantasies do help human beings discharge their hostility.

There are several conditions which apply to the doctoring of psyches with violent fantasies, however, and it may be the case that the Japanese success is due to their recognition of these factors. For televised violence to be cathartic, it must clearly be fantasy. If the violence is thought to be real, its effect will be the opposite of what's wanted, in that it will stir up aggression in the viewer. This was the conclusion of a study Seymour Feshbach mentioned in passing, the one he contributed to the 1972

Report to the Surgeon General. Feshbach had taken two groups of children and shown them the same six-minute televised film of a campus riot. One group was told it was an NBC newsreel of an actual riot, and the other was led to believe the footage came from a Hollywood movie. The aggression levels of those who thought they were seeing real news coverage went up, while the children who felt they were viewing a fantasy experienced decreases. This prompted Feshbach to say during our interview, "When it comes to aggression on television, I think we have to be concerned about the context in which the aggression is depicted. The more realistic it is, the more I worry about it being translated into real life behavior."

It is the fantasy element of televised violence which accounts for its success in removing hostility, as Feshbach stated. Broadcast scenes of real violence impinge upon the viewer's consciousness and create the pressure that might move him in the direction of an actual, matching display of aggression. But fantasy—and this is its glory—makes contact with the deeper, unconscious regions of the mind. It does not challenge the viewer, but rather sets up situations in which he can pretend to be as violent as he might like to be if he were not the subject of social norms.

The Japanese may be on the right track when they keep their violent fantasies quite removed from the realities of their lives. By using foreign fantasies and historical tales, they may have achieved the distance which excludes any carry-over into the real world. Successful fantasies cannot be too far removed or the viewer will not be able to vicariously enter into them, but short of that limitation there are great advantages to displacement.

Another aspect to the utilization of media violence, one which the Japanese may recognize better than we do, is that it only works on those who are in need of it. If the viewer doesn't come to the television set with high levels of constrained aggression, then the fantasy violence can actually raise hostility. This emerged in Feshbach's 1961 study, whose major conclusion was that filmed violence had a cathartic effect upon enraged college students. But for students who were not insulted, and whose hostility levels were therefore low to begin with, the fantasy boxing match elevated their aggressive feelings. The more seething the viewer is, the better fantasy violence will help him discharge his anger; the calmer he is, the more likely it is that fantasy violence will rile him up.

The fact that media violence can agitate tranquil subjects may be another reason that laboratory studies usually indict violent fantasies, to the satisfaction of Media Snobs. Following in the footsteps of Albert Bandura, researcher Margaret Hanratty Thomas (one of the defense's expert witnesses at the trial of Ronnie Zamora, called upon to testify that violence on television can cause violence in the real world) conducted an update of these laboratory experiments in 1974. As before, children exposed to televised violence displayed aggression themselves when given the opportunity. But when *two* children were involved, Thomas discovered, the amount of aggression released was greater still. This led her to think that previous laboratory studies had underrated the amount of rowdiness inspired by media violence. Snobs may find confirmation of their distaste for television in such experiments, but the missing factor that ruins their applicability to the real world is that of viewer choice. The subjects in these experiments are compelled to view, while individuals in the world-at-large turn on violent fantasy only when they have a need for it. When violent television fantasies are freely chosen by someone who needs them, they will do what they are supposed to do and purge violent feelings.

Media Snobbery has had its impact upon public opinion regarding television violence. Instead of being appreciative for the psychological services that the fantasies provide, as the Japanese might be, Americans voice negative judgments. In a 1977 Harris poll, 71 percent of those canvassed agreed that there was too much violence on the medium. Two-thirds of all teenagers concurred with the idea that television action/adventure prompts social violence in a 1978 Gallup survey.

And yet, when investigators probe more deeply, these attitudes turn out to be weakly held at best. A marketing corporation questioned people about television violence in January 1977 and electronically calculated the strength of feeling in their voices. Seventy-eight percent had agreed there were too many violent programs on television, and 63 percent claimed they avoided such shows. But the voice analysis showed that half of those who said yes to the first question and two-thirds of those saying yes to the second did not in actuality care one way or the other. Sixty-seven percent of Gallup's teenagers may have felt that television violence brings on social violence, but the percentage that *opposed* taking these shows off the air was 73. A survey of viewers' concerns was undertaken through Syracuse University's School of Information Studies in 1978, and reservations about violence were far down the list. Most important to viewers was that they continued to get the "predictable plots and repetitive stories" they were used to.

The shallowness of viewers' convictions on television violence may be the upshot of having two conflicting feelings. On the one hand, they have been led by Media Snobs to think of fantasy violence as socially damaging, and on the other they recognize at some level that it is psychologically beneficial. Their desire for these benefits is revealed by the high Nielsen ratings for violence-heavy shows. No matter what people may say, they still want what these programs offer.

That television violence does bring about a beneficial cleansing of hostility-laden minds is indicated by the fact that American fans, when polled, are genuinely unable to recall any violence on their favorite action/adventure shows. A person is aware of violence on the shows he doesn't like so much, but the ones he is fond of do not strike him as violent, despite how vicious the shows reveal themselves to be by objective count of the acts of aggression. This is a significantly revealing delusion, for it signals the fantasies are working. A viewer is receptive to his favorite program, and permits it to enter his unconscious mind as he enters into its fantasy. Operating on a dream-like level, he can vicariously aggress along with the show's hero, and so deplete his own reservoir of repressed animosity. If the viewer *did* recall the show's violence afterward, that would mean the psychic trash removal had not occurred, and the aggression, instead of departing, had become lodged in the mind. It would be as if your garbage man, instead of taking your trash away one day, decided to move in with you. But the inability to recollect the violence means that the fantasy has found and captured the hostility in the viewer's mind—and has left. With the resolution of the program, as the criminals are caught, the viewer can feel untroubled about the aggression he has released along the way, without moving a muscle. There is little sensation in the aftermath except relief.

8

Television Is Good for Your Heart

Soap Operas vs. Prime Time

The instant she pulled into her driveway Linda Miller of Denver, Colorado, knew something was wrong. A truck was parked in front of her house, and to her amazement some of her own furniture was inside, including the television set. Suddenly the truck shot away, its rear doors flapping. Two men dashed out of the house, ran across the lawn, and jumped into the cab of the accelerating truck.

Although her 19-month-old baby was with her, Mrs. Miller took off after the burglars without a moment's hesitation. Blowing her horn, she chased them around the suburbs of Denver, up one street and down the other. "They turned toward me, and I turned toward them," the 27-year-old woman later recounted. "I crossed over to their side of the street and aimed straight for them. I was mad." At the last possible moment the thieves' truck swerved. But it wasn't much further down the road before the three men pulled over to the side; they had decided to negotiate with the irate housewife. When they offered to give back the television, she made them place it inside her car, beside the bawling baby. Appeased, she abandoned the chase, although she did call the police, who soon caught up with the winded thieves.

What had incited the 5-foot-2-inch Mrs. Miller to risk her safety and that of her child, to chase after professional criminals with such ferocity? "Anyone who deprives a woman of her soap operas is asking for trouble," she explained firmly to a reporter.

Mrs. Miller is certainly not alone in her devotion to soap operas. It is a passion shared with tens of millions of other fans, the greater majority being like herself, adult and female. Not only is the audience immense, it is also continuing to expand, as the proportion of total viewing hours given over to soap operas grows while other daytime program types hold even or decline. Two-thirds of all women above the age of 18 now tune in to the serials each week. This includes three-quarters of all women unemployed outside the home, but astoundingly enough, over half of those holding down jobs still manage to sneak in at least one episode during their five days of work. Lady Bird Johnson and Betty Ford, in spite of all their duties, were soap opera fans, and so was Lillian Carter, mother of the President. The audience is not exclusively female; the 20 percent that aren't include Joe Namath and Sammy Davis Jr. Representative Anthony Moffett (D-Conn.) revealed that, around midday, many of his fellow Congressmen retire to the cloak rooms to catch the latest developments in what humorist James Thurber, writing a series of articles for the *New Yorker* some years ago, called Soapland.

When historians look back on post-World War II America from the vantage point of decades and centuries ahead, what may strike them as conspicuous about this era is the time and attention spent with the unique cultural creation of soap operas. It is not a preoccupation that has caught the attention of many behavioral scientists and social observers so far, but it is one that will stand out in high relief in a long-range perspective. At least one episode per week is watched in 40 percent of all American households, for a total weekly audience of over 50 million human beings.

While the antecedents of soap operas can be traced back to earlier serial forms, to the *Iliad,* the *Arabian Nights,* Daniel DeFoe, or Charles Dickens, the full-blown genre didn't appear until broadcasting came along. From inauspicious beginnings on Chicago radio stations in the 1930s, it has become a dominant art form, available throughout the English-speaking world as well as Europe, Asia, and South America. Like jazz, it is a major American cultural invention and export. More of the world's women participate in soap opera viewing than in any other single cultural activity. For them it is the premier dramatic experience, the one that holds them as no other can.

Like situation comedies and action/adventure shows, soap operas too are fantasies, but in most other respects they have little in common with their prime-time cousins. Very little occurs under the viewer's gaze which can be counted as action. Plenty is reported, but most of it takes place offstage. There are few depicted car chases, shootings, crashes, falls, dashes, tumbles, or anything else remotely kinetic and chancy. Although the 1970s saw more on-location production, still the standard pattern remained: whatever transpires largely does so indoors and in words. Communications scientist Natan Katzman closely analyzed one week's worth of content for 14 soap operas and reported that of 884 settings studied, only nine were not indoors. The expansiveness of prime-time shows, the outbursts of laughter and aggression, are largely lacking in the interior dramas of the daylight hours.

The characters too differ markedly from their evening cohorts. In fantasies broadcast during prime time, women are scarce; by count there will be four male characters to each female. But in soap operas parity has been achieved, and the number of women equals the number of men. These women are of a sturdier sort than the slight, sexy lasses who usually replace them after the sun goes down. Many will be widowed, divorced, or separated—people making their own way through the world, although with no shortage of lovers and friends. Concerned with their own well-being as well as with the well-being of others, they manage to roll with the punches. Some of the punches are thrown by another character type not come upon so often in prime time—the evil woman, the femme fatale, out to undermine the happiness of everyone else.

Male characters are serious, well-kempt professionals—doctors and lawyers often. The camera captures them in book-lined offices more than their evening colleagues. A sharp contrast is that there is little of the crime-fighter to the daytime males. In fact, there is little heroic about them in any way, and eventually they may show themselves to be weak and inadequate to the women in their lives. Action/adventure heroes never fail, but these men are liable to.

The settings are as dissimilar as the characters. Prime-time fantasies are often urban, but soap operas are generally located in a small town about 50 miles from New York, Chicago, or Los Angeles. Over the years the town may have grown some, to incorporate the offices, hospitals, court rooms, and other locales necessary for an up-to-date soap opera, but it retains much of its small town feeling. For the most part it is a sustaining, well-mannered community, although like a magnet it seems to hold an irresistible attraction for evil and catastrophe.

The problems that propel soap operas are not like the ones viewed later in the day. There the problems emanate out of the characters, their foibles or larceny. Here the problems often simply befall people. Without the characters deserving it or inviting it, calamity strikes. The unexpected, the unknown, is the spur to the story, making bad luck a prime theme. Much of this bad luck takes the form of unwanted pregnancy. A chance encounter or a rape, and someone is with child. Soap opera women suffer a birthrate that is eight times as high as the United States birthrate, and higher than that of any underdeveloped country in the world.

While Katzman was doing his week-long analysis of soap operas, he made a list of the calamities he observed:

Romance and marital affairs:
>3 romances in trouble
>4 marriages in trouble
>8 clear cases of marital infidelity
>2 cases of potential marital infidelity
>3 divorces or annulments

Medical developments:
>2 mental illnesses
>4 psychosomatic illnesses
>5 cases of physical disability
>4 pregnancies

Social problems:
>3 cases of business difficulties
>3 professional men on probation or fired
>2 cases of drunkenness
>4 youths involved with drugs
>4 offspring of unmarried parents
>5 cases of family estrangement

Criminal and undesirable activity:
  blackmail
  bigamy
  3 threats
  2 murders
  2 other deaths
  poison
  drug traffic

"On the whole, the world of soap operas was full of troubles," he concluded with scholarly understatement.

Soap operas are not only daytime, they are also daily. A prime-time show comes but once a week, but soap operas are there day in and day out, a constant presence in the viewer's life. In truth, the average viewer sees her favorite show only three times a week, so to oblige her there must be a certain amount of repetition and restatement. Repetition is also required because each soap opera consists of several story lines, and when one of these threads gets picked up anew, the audience has to be reminded what it is about.

The most profound difference between the two kinds of programming is, like most profundities, everywhere felt and everywhere obscure, lying deep below the surface. It has to do with dramatic time. In soap operas time is felt to pass. The stories evolve, relationships change, personalities mature or take a turn for the worse. If the soaps are repetitious, it is because there is something to repeat. But time stands still in situation comedies and action/adventure shows. There roles are rigidly set and relationships are static. Kojak does not develop amnesia, drinking problems, or impotency, and Mork does not turn on Mindy. Each prime-time episode may entail some small passage of time, but when we tune in next week, it's back to the starting point again.

The flow of time is assured in Soapland by the absence of climaxes. Being unlike prime-time dramas and having no resolutions, soaps have no stopping points. Subplots may end, but the soap opera itself continues on and on. Old characters fade off while new ones, with new troubles and new tales to tell, are knit into the endless tapestry. If each broadcast were entire unto itself, and ended with dramatic closure, then they would resemble the evening shows, where time is held back.

All these differences point to the fact that soap operas are created to serve quite other psychological purposes than prime-time fantasies do.

Production

Even the production of soap operas differs extensively from the production of prime-time programs. Situation comedies and action/ adventure shows are mostly concocted in Hollywood, while New York City is the real-world home of Soapland. One reason for the existence of soap operas, someone once said flippantly, was to keep the large crops of otherwise meagerly employed New York actors and actresses in paychecks. The difference in sites reflects a difference in ownership: prime-time series most often belong to a West Coast production company, but soap operas are owned either by the network or, harking back to the early days of broadcasting, by the sponsor. Procter and Gamble still directly controls the production of several.

In the beginning, during the time of the freeze on new stations, no one was absolutely sure that soap operas could be produced on television. Radio, where they had been thriving for fifteen years, looked to be the preferable medium for several reasons: whatever was referred to in the scripts did not actually have to exist as props and backdrops; the actors did not have to commit their lines to memory every day; the imaginations of the audience could have freer play. Television would not do, skeptics thought, because the difficulties of production would result in something so deflated as to be unenjoyable. And in fact, the first network soap, CBS's *The First Hundred Years,* died within one season from viewer uninterest, providing merriment for those who seized on its name and fate. No radio serial successfully made the transition to the newer medium, with the exception of *The Guiding Light,* which did it only by running concurrently on radio and television for five years.

Nevertheless, the problems were surmounted, and soap operas quickly came to dominate daytime television. Seemingly little was lost as scenes became exclusively interior, of the sort that could be constructed cheaply in a studio. And actors rose to their challenge through a combination of resilient short-term memories, stereotyped lines, and the innovation of teleprompters. The economics of production forced a lengthening of program time from radio's fifteen-minute show to a half hour, since one thirty-minute episode was far cheaper to produce than two fifteen-minute ones. By 1960 no more radio soaps were being broadcast. If individual shows did not make the transition to television, the genre had made it—and totally.

The major incentive driving the television production of soap operas, before which all obstacles fell, was profit. Manufacturers of products which a housewife might purchase knew that daytime programming was a sure way to put their goods under her nose. They were determined to see the production of shows which would capture her attention. The networks were if anything more resolute because the financial rewards for them were even more immediate and clear-cut; they didn't have to wait for people to make their way to markets and purchase a floor cleanser, nor to untangle the complicated chain of events that link commercials to purchases. Their business was simpler: so many viewers translated to so many dollars.

Over the years the payoffs for the networks, and presumably for the advertisers, have remained high. Soap operas may have fewer viewers than prime-time shows, but production costs are far lower and allowable time to be sold for commercials is mc.e ample, with the result that daytime television is even more profitable than prime time. A week of an hour-long soap opera can cost up to $200,000 to produce, but the advertising yield from the 30-second spots sold at $15,000 each can easily reach $600,000. Fred Silverman, whose first network job was in daytime programming for CBS, may not have begun in the most prestigious segment of the television business, but it was the most profitable, and thus the most important in terms of the lifeblood of the industry.

Among the ways that soap opera production differs from prime-time production is that the kingpins are not the same. In creating programs destined for the evening hours, the central figure is the producer. He is the man who has all the reins in his hands, who maintains communication with the network, handles the finances, hires and oversees the staff (including the director, who as different from a movie director is more restricted to technical execution), works with the writers, supervises casting, and so on. ABC's Bob Shanks says, "In television the producer is certainly the commander of the show. Ultimately, he is responsible for every detail." A soap opera producer, on the other hand, may well have all these duties, but has surrendered a great deal of control to those considered to be the mainsprings of the production—the writers. Soap operas are above all a writer's product. This is suggested by the name for the long-range story outline which prescribes all the ins and outs of the plot to come: it's the "bible."

The head writer, who may earn a one million dollar annual salary, constructs the bible for six or 12 months ahead. Every liaison, every bout of amnesia, every disappearance, every surgical operation, is put down in writing. At a story conference she will review the bible with the producer, director, and others responsible for the production, but the bible is primarily hers. Then, based on this outline, the head writer together with other writers will construct sketches, called "breakdowns," of each scene for the next several weeks ahead. Oddly, the actors' lines are not written by these writers. They are done by assistants whose sole job is to create the dialogue within the dictates of the breakdown and the allotted broadcast time. This happens quite close to the deadline, and the actors don't see the script until a day or two before taping. There wouldn't be much point to putting it in their hands earlier; they're too busy learning that very day's lines.

A year's production of a prime-time show can consist of 26 episodes, but a soap opera is shot five days a week, year round, for an astounding total of 260 shows. Every one of those days the production routine is the same. Having studied their scripts the evening before, the actors begin to assemble at what is breakfast-time for most other working Americans. In a bare rehearsal hall they walk through the episode to be taped, prompting themselves from the typed dialogues in their hands. Then hair, makeup, and wardrobe are attended to before the troop moves to the studio for camera blocking, which in effect is a rehearsal for the production crew. Still holding their scripts, the actors have a run-through for timing, checked by assistants with stopwatches. Lunch follows, and then a complete dress rehearsal. It is late in the afternoon before the show is actually taped. Once the tape is approved, the actors are free to go home, where they will spend time going over their lines for the next day.

The daily production of an established soap opera usually goes smoothly, so smoothly that the routines will not be tampered with for years at a time. Writers, actors, and crew learn to mesh and work well with each other, in the interest of putting out a series that is ceaselessly appreciated by its audience. As Edwin Vane, a vice-president of ABC, remarks about soap operas, "When a serial works, it has longevity. Once you wind it up, it's good for the long haul—with care and feeding." Longevity is another characteristic of daytime serials not shared with their prime-time relations. Even the most popular evening shows are unlikely to last more than seven years, while in a recent tally only two of the 14 televised soaps were younger than 10 years, and three were over 25.

The longest-running soap opera of them all, first aired in the fall of 1951, is another of the daytime fantasies carefully tended by the production arm of Procter and Gamble. The title chosen then, *Search for Tomorrow,* remains perfect for a daily drama. Along with its predecessor, *The First Hundred Years,* and its companion over the decades, *The Guiding Light,* the title suggests the relentless groping forward that most viewers sense is an analog of their own lives. The title conveys the idea of negotiating one's way past what time brings, which is the feeling many successful soap operas transmit.

Search for Tomorrow is set in the mythical town of Henderson, somewhere near, but not too near, Chicago. Since real-world Chicago is where radio soaps began, this television one has stayed close to the source. The central character is the off-and-on widow, Joanne Barron Tate Reynolds Vincente. According to Roy Winsor, creator of *Search,* "We did get the idea for *Search* from radio. I wrote the original presentation for the serial and saw Joanne as a kind of young Ma Perkins, the sort of woman who cared about her neighbor's problems, who would offer help to others, and who would face her own personal problems with dignity."

Like most soap operas *Search* has become more "layered" as time has passed, in that more subplots (especially those thought to be appealing to younger viewers) have been folded in, and so Joanne has had to share the spotlight with other characters. Yet the kindly, long-suffering Jo, and the story of her ventures and marriages, remains at the hub of the serial. One after the other, her husbands have managed to let her down. The first, Keith Barron, got himself killed in an automobile accident. Number 2, Arthur Tate, turned into a weakling and soon died of a heart attack. A lengthy romance with Sam Reynolds ended unhappily when he was dispatched by the UN on a peace-keeping mission to Africa, got lost, and finally returned a changed man, cruel and jealous. He had good reason to be jealous: Jo was taken with the handsome Dr. Tony Vincente, who saved her eyesight after a spell of blindness. She married the good doctor, but he too succumbed to a heart attack. Despite all these tribulations, Jo has stayed a person of sterling character, always ready to face the day.

Perhaps the most remarkable feature of Joanne is that, since the first episode aired on September 3, 1951, she has been played by the same actress. Mary Stuart has appeared on more hours of television than any other performer ever; three generations of women have viewed her loyally and intensely. At a cast party celebrating the 25th anniversary of the first broadcast, held September 4, 1976, the actress was sung a light-hearted song by those connected with the production (to the tune of the "Battle Hymn of the Republic"):

> With broken heart and body she has managed to survive.
> The ratings go on soaring just as long as she's alive.
> So keep her paraplegic for another 25,
> And the show keeps rolling on.
> Glory, glory, Mary Stuart,
> Glory, glory, Mary Stuart,
> Glory, glory, Mary Stuart,
> The show keeps going on.

Mary Stuart may be assured of a job with *Search* as long as she desires, but most soap opera performers are not so fortunate. While the standard acting contract is for three years, the producers can release an actor at any of the thirteeen-week marks. It was the search for higher ratings which caused actor Kenneth Harvey, after seven years with *Search,* to be written out of the show in 1975. A revision of the bible specified that he would suffer an automobile accident and subsequent brain damage and be operated on by Joanne's husband, Dr. Tony Vincente. Paralysis would occur. While lying comatose in his hospital bed, he would have the plug pulled by an unseen murderer.

As the day for taping the murder drew closer, the actor noticed changes around the set. "I observed that people had begun to perform odd little rituals more proper to sickrooms and funeral chapels than to a television studio. Stagehands would tiptoe their way carefully around me, turning their eyes away from where I was lying. I would find myself, at odd times, the recipient of gentle pats on the back, sudden silent embraces, impulsive kisses on the cheek, and wide, sympathetic stares from across the room." Then the deed was swiftly done, the character expired, and the actor was let go.

Not all soap opera deaths are handled so sensitively as they are on *Search.* A letter-writer to a soap opera fan magazine wanted to know the fate of Lee Baldwin on *General Hospital.* The character had married widow Caroline, who had son Bobby of the rare nerve disease, who was engaged to nurse Samantha. One day the four of them were simply not on the show; no mention was ever made again. What happened, the viewer wanted to know.

Well, answered the editors of *Soap Opera Digest,* the four had gone to

Florida, been in a boating accident, and their bodies had sunk without a trace.

Not quite. Jean Kerr, the writer, was also a fan of *General Hospital,* and also wondered about the fate of Lee Baldwin and clan. One evening at a New York theater intermission she bumped into the actress who had played Caroline. Soap opera fans have a hard time distinguishing actresses from their characters; for all her fame and sophistication, Miss Kerr was no different. "Caroline," she blurted out, "what happened to all of you?"

The deposed soap opera star replied slowly and grimly, "Mr. Silverman didn't like us."

Adding to Minds

Let's stay with the fact that soap opera fans will confuse the actress and the character. Mary Stuart tells this story: "They insist they know you. I had a lady come up to me in Bloomingdale's one day and she said, 'You never wrote me a thank-you note.' And I said, 'For what, madam?' She said 'You came to my house for a party. You never even wrote me a thank-you note.'"

Bloomingdale's, the fashionable, racy New York City department store, seems to be a likely meeting-ground for these confrontations. Eileen Fulton, who plays the evilest of all soap opera femme fatales, Lisa Shea on *As the World Turns,* was slapped by an outraged viewer while shopping at Bloomingdale's. On another occasion Miss Fulton was in the furniture section of the large store just when *As the World Turns* came on the air and overheard women customers saying how much they hated Lisa and wanted to kill her; the actress fled.

Peculiar happenings stem from the inability of some viewers to distinguish between the performance and the private lives of daytime actresses and actors. Eileen Fulton's father is a Methodist minister, and when Lisa Shea was in the middle of a torrid romance with a character named Bruce, one of his parishioners wrote him a note to say, "We saw what your daughter did in the bushes with Bruce." Don Hastings, also an actor on *As the World Turns,* playing Dr. Bob Hughes, says that he was once asked to take over a medical practice in the Midwest. But nothing could be quite so bewildering as the letter to an actress who had been playing the part of a Broadway star on a popular soap. The letter praised the performance at some length before ending, "Have you ever thought of becoming an actress in real life?"

Other performers in other sorts of popular arts don't suffer this confusion to the same extent. When actress Emily McLaughlin of *General Hospital* walks down the street with her husband, movie star Jeff Hunter, people will address her as her character Jessie Brewer, and will ignore her husband. What's going on here? Miss McLaughlin believes, "We are in their living rooms five days a week, leading a continuous life, so we achieve a kind of reality. That's why viewers think we're real people."

Many viewers do genuinely experience soap opera characters as perfectly real. Whenever a favorite gets "married," the network will be flooded with congratulatory cards and letters, and when another "dies," telegrams and sympathy cards pour in. Thousands upon thousands of letters warn the heroine of impending dangers, give straight-from-the-shoulder advice, praise good behavior, goad action. Telephone calls relay reproofs, cautions, schemes, pledges of friendship. The soap opera producers and writers are delivered bundles of mail which ask the "lawyers" for legal advice and the "doctors" for medical consultations. According to CBS, poor characters on their soaps had to be eliminated because too many CARE packages kept arriving.

Soap opera fantasies, unlike other fantasies on television and elsewhere, are often not felt by their audience to be divorced from the real world. Instead they are tightly, near indivisibly, attached to the viewer's sense of real life. The threads of the two fabrics are woven together. Cinema critic Renata Adler, confessing her addiction to the soap opera *Another World,* related, "When Lee Randolph died, a suicide who had lingered on for weeks, I watched her face being covered by a sheet, and I was ridden by the event. But it was not at all like losing a character in fiction of any other kind. I saw the characters in the soaps more often than my friends. It had a continuity stronger than the news."

What soap operas do for Renata Adler and Jean Kerr, together with millions of other loyal viewers, is to add to their personal social universe—extending it, peopling it, marbling it with developments and events. Made social creatures by our genetic inheritance, all human beings have a capacity for emotional and cultural participation in the larger group, a capacity which modern life, with its isolation and anonymity, does not always fulfill. This can be especially true for the homebound, who may be detached from any community, removed from the workplace, and during daylight hours devoid of family too. They want their private world widened, and this is exactly what soap operas accomplish for them. The employed have less need for these fantasies because they are compelled to enter into so much interpersonal contact that their social capacity may be exceeded. However, for those who stay at home, their world can be a curtailed one, and their needs for extension and amplification can be high.

An advantage of soap operas is that they augment private lives at little cost to individuals. Empty spaces in minds are filled, and people are given something to think about, at no emotional risk to themselves. To achieve the same sense of social fullness in the real world would call for considerable effort and no little psychological expense.

So deft are soap operas at augmenting lives that they can become, to some extent, the social universe of choice. This is why women who go to work may still elect to view the daytime serials. For them as well as for women who remain at home, the shows displace some of their less satisfying social contacts. Writing up his examination of soap operas, Natan Katzman said, "Soap opera characters have replaced neighbors as topics of gossip. To some extent the programs have replaced gossip itself." Not only do soaps displace local reality, but they can also be preferred over crucial information from the world-at-large. When Ronald Reagan and then Pope John Paul II were shot in 1981, many soap fans complained about the coverage interfering with their favorite shows. One station executive reported, "They harassed the station. Some called repeatedly. One threatened to come down and do bodily harm to the people at the station unless the Pope story went off the air by 1:30."

The great thirst of the audience for this augmentation is revealed by the existence of soap opera fan magazines, whose function, over and above performer and character profiles, is to recount the recent plot twists for addicts who for one reason or another are unable to keep up. According to Paul Dennis, editor of *Daytime TV,* "New magazines on the soaps keep coming out and we're so afraid that we'll all burn ourselves out. But the strange thing is that no matter how many new magazines come out on the newsstands, the circulations keep going up, not down, pointing out to me that there's really no limit to the interest people have in the subject." No other genre of television is backstopped by such publications because no other audience cares so avidly about what happens. But soap opera viewers, should they miss an episode or two, will do whatever it takes to find out the story line. They are afraid if they miss a key segment, the continuity they want to vicariously participate in will be lost. So plot summaries and details are pursued as for no other television fantasies.

The way soap operas engineer this high viewer involvement is by not having resolutions at the end of each episode. This turns out to be the most significant difference, in terms of the dramas' structures and effects, between Soapland and the climax-capped prime-time programs. Without climaxes, the soap opera viewer is kept perpetually on tenterhooks, forever wanting to know what happens next. The plots and characters continue to writhe in the viewer's mind, and the result is that her social universe becomes that much more enlarged. No dramatic resolution means no ridding oneself of the story.

This lack of completion is the melodramatic device which accounted for the largest audience ever collected for a television serial episode. Over the years a few soap operas have managed to locate themselves in the alien territory of evening hours where the potential viewership dwarfs the daytime crowd: *Peyton Place* was one such show in the mid-1960s, and *Dallas* in the late '70s and early '80s was another. In the final segment of the 1979–80 viewing season, the writers of *Dallas* had the magnificently baleful lead character, J.R. Ewing, shot by an undisclosed hand. Through months of summer reruns and a delaying actors' strike in the fall, a swelling number of Americans queried one another about the identity of the assailant. When the sister-in-law's crime was finally revealed in November 1980, over three-quarters of the television sets on in the United States were tuned into the CBS broadcast.

But in the usual prime-time fantasy the story does not carry over from one telecast to the next. At the end of each episode all the loose ends will be tied up and the viewer will experience closure. Resolution will bring the viewer a sense of relief, however minor. The fantasy will not dwell on in the mind; its job is done and it departs, taking with it what tensions and mental debris it has managed to wipe up.

In a nutshell, the fundamental reason soap operas stand in contrast to typical prime time shows is that the psychological services provided by these two types of television fantasy are as different as day from night. Soap operas add to consciousness, contributing a scaffolding that viewers can use to strengthen and extend their own lives; prime-time comedies and adventures subtract from the unconscious, removing stress and strain stored there. This is the reason that soap operas can be recited in the minutest detail by their viewers, while prime-time shows will be recalled imperfectly, if at all. It occurs to me that heavy viewers of daytime fantasies may well need a dose of the evening variety to reduce some of the burden of Soapland.

So real are soap operas to their fans that psychotherapists are finding them to be a convenient point of entry into the minds of troubled patients. Some patients are highly resistant to discussing the real horrors of their private lives, but can begin psychotherapy by referring to the soap opera characters that populate the outer margins of their consciousness. Psychiatric social worker Anne Kilguss has written about using this approach with success. "From the program, one works back to the individual and her concerns," she says. One reluctant patient heard mention of a soap opera abortion, volunteered that she too had undergone an abortion, and so began her course of therapy. Other psychotherapists report using soap operas in group therapy sessions, where all patients can view the same soap episode and then use their perceptions as a jumping-off point for discussing their own feelings and behavior.

Media Snobs have a hard time conceiving of soap operas as positive, therapeutic adjuncts to reality. It is the unreality of daytime serials that they point their fingers at. They say they are perplexed by the onslaught of bizarre illnesses and cracked romances. Feminist Snobs insist women are being caricatured on the shows—their roles aren't brazen or professional enough. And Rose Goldsen, referring to the "mass desensitization sessions which soap operas target to the nation's homes," claims the daily broadcasts are teaching viewers to discredit family ties. Soap operas, she says, "whittle away at the fundamental sense of trust every human family tries to imprint on its members. The job is subtle and effective, like water wearing away rock."

It is true that daytime fantasies, unlike other television fantasies, do teach things. Akin to real life, they can be just as instructive. When the information that is conveyed is of an especially laudable sort, soap opera producers will pat themselves on the back. For instance, by repeatedly mentioning Pap smear tests over several months *The Guiding Light* greatly increased the public's awareness of this simple test for cervical cancer. But most of the information exchanged is of a more emotional sort. In 1943 behavioral scientist Herta Herzog asked housewives what they gained from listening to radio soap operas, and learned that advice on personal problems was one of the major attractions; it is still true today with television soaps. Daytime serials, concludes Columbia University professor Herbert Gans, "have supplied housewives with information about how to solve their own problems; even if the problems the soap opera characters have often seem sensational or exotic ones not often encountered by their audience, they have provided the message that people, as individuals, could solve their own emotional and social problems through their own efforts and the right kind of information."

In contradiction to what Rose Goldsen argues, a mistrust of relationships is not the chief lesson of the soaps. Just the opposite. When interviewed fans will say that the serials teach them "how to be with a husband," "how to handle problems with children," "how to have a romance." What they are learning is how to manage their close relationships. Soaps put a strong, positive value on human ties, and drive this home to their audience. On *Search for Tomorrow* the only other characters to endure from the program's beginnings in the '50s were Joanne's friends the Bergmans, the couple she could count on to be close by with a smile or a helping hand. Stu Bergman has always been played by actor Larry Haines, and until she died in 1971, actress Melba Rae played Marge Bergman. So wrenching was the actress' death to the show that she never was replaced, and the writers allowed Stu to become a widower. Ties like the Bergmans with Jo are the core of soap operas. At center it is matters of the heart that daytime dramas deal in, for this is the brand of human activity that viewers want most to explore, sense, and receive instruction about. If the soaps look at human relationships in all lights, it is to get the best out of them, not the worst.

The bonds between humans are highlighted in the trying crises that fate relentlessly and indecently brings. As in life, fate is the true villain of Soapland—a point made over and over by the deaths of good characters and the inflictions imposed on happy families. The greater the misadventure, the more cherished the personal ties become, since they are the only solace and the only avenue toward peace and satisfaction.

The lesson the soaps have to offer their viewers is that life's blows can be survived, and that fellow-feeling is the best cushion. By one means or another we'll get through. Survival is the baseline moral of daytime dramas, gained by keeping eyes fixed on the higher, more elusive goal of happiness. The dialogue writers had one soap heroine say, "Oh, life does work out so wonderfully sometimes. If people could just have patience— if they could just get through the hard times and know that around the corner there may be the most glowingly wonderful time waiting for them." This is what every viewer wants to know.

Fate, affection, survival, the chance of happiness—these are the basics of real life, and the elements of soap operas. Soapland is a graft on to the real world. It is, writer Judith Viorst once said, "life with most of the dull parts deleted." For viewers who become citizens of Soapland, their own lives become fuller and better informed, as well as better managed. In the words of a young man who explained his soap opera viewing to a reporter from the *New York Times,* "When I'm watching a story, I always put myself in the position of a teenager and figure out how I would handle some of the situations and what I'd say to my mother. I don't think my life would ever end up like the soap operas on TV, but everyone's life is a soap opera to themselves."

9

Television Is Good for Your Brain

Information Received

A segment of CBS's *As the World Turns* was being seen early one Friday afternoon when without warning it was interrupted by another kind of television content, a news bulletin. President Kennedy had been shot in Dallas. At the time CBS had no studio facilities for unplanned newsbreaks, so the tragic story first came in the form of Walter Cronkite's voice alone, strong as ever, but touched by the differing feelings of melancholy and alarm that were rapidly to fan across the nation.

The soap opera resumed, to be interrupted again by another audio report from Cronkite, confirming the first. Twenty minutes slipped by before furious work by network technicians could displace the afternoon dramas and unite Cronkite's image with the audio signal. Reality had to strain to overcome fantasy. At 2:33 P.M., when the official announcement of Kennedy's death was relayed, viewers could clearly see the tears pooling in Cronkite's eyes.

Decisions were immediately made in network headquarters to suspend entertainment programming and commercials, and for the next four days the public viewed nothing but the details of the assassination and its aftermath. American households were tuned to the coverage for an average of 32 hours during the long weekend of November 22–25, 1963. Millions upon millions gasped at the killing of Lee Harvey Oswald on Sunday and wept as the President was buried Monday. Their grief bore down on them when they saw the Kennedy children—little John copying a salute, Caroline following her mother to bend and kiss the flag on her father's coffin. Never before or since has such a common and intense reality been shared in by so many people for such long stretches of time.

It was due to television that Americans learned about the assassination so rapidly and accurately. That a story could travel any faster through a nation of 200 million individuals can scarcely be imagined. According to follow-up research studies, nine out of ten citizens knew of Kennedy's death within an hour of its being announced. About half learned directly from the broadcast media, and the rest got the news second-hand from friends and co-workers. Television was preferred as a source above radio whenever there was a choice, people told communications researchers William Mindak and Gerald Hursh, because it brought them closer to the event. "The reasons people spent much time watching television," Mindak and Hursh related on the basis of interviews conducted the week after the event, "are to be found in the uniqueness of the coverage. Neither radio nor newspapers could match television's facility for realism and psychological proximity." Because the air was thick with rumors (it was said that a Secret Service man had done it, that Jacqueline Kennedy was dead too, that Lyndon Johnson had been shot), people needed to see for themselves what was really going on. The networks had the pictures they craved.

During the following week when things could be checked, the accuracy of television's surveillance of the events in Dallas was conceded by most viewers. Few of Mindak and Hursh's respondents felt the networks had failed them in capturing and transmitting all the known details.

As the coverage had proceeded that weekend, the news accounts had taken on a second function. Viewers reported later that television helped them to absorb the blows and look beyond the tragedy to better times.

"It gave timely reassurance by showing the existence and continuity of cherished institutions and values," deduced Mindak and Hursh. Walter Cronkite and the other newscasters, together with the people they interviewed, made the audience feel that things were not going to fly apart, that the guiding values of American life were still in place. "After the assassination," said Mindak and Hursh, "television most conspicuously contributed to the enforcement and affirmation of social norms and values. Faith in the future of the country, belief in the form and continuity of the government, these are the cornerstones of the American value system. The comments of 92 percent of those interviewed show that television reawakened these values for them."

There is much to be learned about the broadcasting of news and information from an inspection of the Kennedy assassination coverage. It is a repulsive truth, but it is the case that the murder of a President is the biggest news story of all. "We cover the President expecting he will die," confessed a Washington bureau chief. No other news account could be more horrifying for Americans or hold their attention more completely. By examining this archetypal story we can see what it is that viewers ask for and get from television news.

First, the public tacitly demands that each and every day news gatherers will be in a position to monitor the real world for whatever is menacing or unexpected. No one was amazed that the networks happened to have correspondents in Dallas covering the President; it would have been odd if it were otherwise. When anything does occur to threaten the continuities and expectations that govern American life, television news is supposed to promptly and accurately inform the public and relay pictures; thus the report of the assassination was immediately broadcast. (In many other countries it would not necessarily happen this way; news of the slaying of a political leader might have to be carefully managed.)

Secondly, once the extent of a threat has been conveyed, then it's the task of the news to help in the process of adjustment. This is what correspondents were doing in the hours and days following Kennedy's death. It is not surprising that Sander Vanocur, as he was flying into Dallas to assist with the coverage and before he knew many of the details of the shooting, was already telling a fellow reporter that he had great faith in the power of the presidency and that an orderly continuation of government under Lyndon Johnson was certain; he was soon to convey the same message to an uneasy nation. In broadest terms, what viewers want to get from their newscaster, whenever possible, is reassurance. Av Westin, the long-time executive producer of ABC Evening News, has remarked that the average viewer wants to learn affirmatively two things: is my home safe, and is the world safe, for the next 24 hours?

The twin duties carried out by the news in behalf of the audience—surveillance and reassurance—were touched on by Senator William Proxmire of Wisconsin in his praise of the reports from Dallas and Washington: "Not only was the coverage dignified and immaculate in taste, it was remarkably competent and frequently it soared with imaginative, if tragic, beauty. The intelligence and sensitivity of the commentary and continuously expressed dedication to this country's strength and solidarity in its hour of terrible grief were superb."

The news audience's need for surveillance and reassurance was objectively confirmed years later in one of the very few studies to explore what viewers get from newscasters. In the mid-1970s Mark Robert Levy, doing doctoral research at Columbia University, conducted lengthy interviews with 240 adults who watched at least one newscast weekly. The services they said television news provided for them could be clustered primarily under the heading Surveillance-Reassurance, Levy concluded from his statistical analysis of their responses. "While viewers use TV news to survey the external environment," he wrote, "they also seek reassurance that the world, both near and far, is safe, secure, and despite the crisis nature of many news items, relatively unchanging, demanding no immediate action on their part." This need for reassurance may be the reason that a 1981 study of 1,000 evening-news viewers found they strongly preferred a network anchor who was "a soothing, comfortable sort." The researcher in charge, Gerald Goldhaber, explained, "The viewer has had a hard day. He wants to hear that things are going to be OK, or at least manageable, while he has dinner."

One of the things Levy noticed about the audience for news is that older viewers are almost twice as likely as younger ones to watch every day. It is logical that, if the news deals in reassurance, those most in need of it will be the most prone to regular viewing. Older people can feel themselves to be at a precarious and vulnerable stage of their lives, and so are drawn to the comfort of the newscasts. Audience figures tell of a pronounced skew in this direction, with the median age of news viewers being much higher than the median age of the American population.

The success of the Kennedy assassination coverage marked a turning point in the evolution of television news. Just two months earlier the networks had lengthened their evening news shows from 15 minutes to half an hour, and it remained to be seen if the expansion of the news divisions was worth the effort. But not long after that momentous weekend polls related that television had become the preferred news medium for Americans. The public's first choice had shifted to the national news that came from television above that brought them by radio, newsmagazines, or newspapers (which did retain leadership for local news). The Bower poll of viewer attitudes in 1970 determined, "In answer to our survey questions about the four media we find television surging ahead of newspapers during the decade as the news medium that 'gives the most coverage,' overtaking radio in bringing 'the latest news most quickly,' edging out newspapers in 'presenting the fairest, most unbiased news,' and increasing its lead as the medium that 'gives the clearest understanding of candidates and issues in national elections.'"

Perhaps because of the rich visual information on the network newscasts, the majority of Americans have now also arrived at the opinion that television news is the most complete and most believable. Asked which medium they would trust in a hard-to-understand or controversial news situation, the respondents in a 1981 survey taken for the *Washington Post* mentioned television first, giving it twice the vote of the second runner, national newsmagazines. In a special study done by Louis Harris for the Senate Subcommittee on Intergovernmental Relations, of all institutions network news was found to have made by far the greatest gains in public confidence since 1963, overtaking the military, organized religion, the Supreme Court, the U.S. Senate, the House of Representatives, and the Presidency.

Television news has done so well at supplying surveillance and reassurance that not only has it emerged as a standout in public opinion, it has also become dominant in terms of actual behavior. That is, as well as claiming they prefer network news, people demonstrably do attend to it more than any other national news source. In this instance actions speak as loudly as words; by objective measure Americans now get most of their news about the world from their sets.

But before broadcasting executives can get too swell-headed about all this, there's the other, less pleasing side of the coin. Television news goes unwatched by most people most of the time. By looking closely at the Kennedy assassination coverage we can learn what kinds of messages viewers want from news. We also learn of the growing success of network news since then at providing those messages—success so considerable that television has rapidly outstripped other news media of much longer standing. What we don't learn is the extent of audience interest under normal conditions. Typically, the nation's appetite for surveillance information and subsequent reassurance is a very small one.

In one survey less than 10 percent of a national sample of Americans mentioned news and information as a reason for watching television. Viewing diaries kept by 6,800 individuals for the W.R. Simmons research firm in 1971 disclosed that over half of the adult population did not watch a single news show during the two-week test period. Many members of the circumscribed news audience stumble upon the news accidently, just because their sets are going; only one-sixth of the viewers go out of their way to switch on the television when the clock tells them the news is about to be broadcast. In 1980 the total audience for all news shows, network as well as cable, usually amounted to less than a quarter of the adult populations.

Not only is the audience for national news relatively small, it is also flagrantly inattentive. Just a third of the initial viewers watch the telecast all the way through, while the rest are diverted, most often for dinner. Many eat and view out of the corners of their eyes; when Levy asked his 240 news viewers, "What else do you do while watching the news?" he learned that a full 40 percent were engaged in their evening meal. With regard to how intensely different program types are viewed when they are on, the 1972 Report to the Surgeon General related that news programs are at the bottom of the list, while entertainment programs are at the top. One researcher telephoned a sample of people immediately after the evening newscasts were over and asked those who said they had watched how many items they could recall; half of them could remember *no* stories from the just-completed programs. The nonchalance of regular news viewers was indicated by the response to a question Levy put them. "If you couldn't see the TV news for several weeks, how would you feel?" he asked, and a third replied that they wouldn't miss it at all.

The very few stories that do register with news viewers are frequently misinterpreted, as a 1980 study commissioned by the American Association of Advertising Agencies revealed. People may see a report on the successful resettlement of Vietnamese boat people in the United States and, when subsequently asked about it, say the story concerned reasons why the Vietnamese shouldn't be allowed to relocate here. Gerhardt Wiebe could easily interpret this in terms of his scheme of *directive, maintenance,* and *restorative* mass media content: human beings tend to turn *directive* messages (content with new information calling for new responses) into *maintenance* ones (messages which review or embellish or elaborate what people already feel to be the case). No one wants to be repeatedly thrown off balance by a stream of *directive* messages; it's much less of a strain if they can be converted into the *maintenance* sort.

This penchant of the news audience to reinterpret new information as confirmation of old views is most plaintively apparent in the instance of political coverage. Early in the television era the social scientist Joseph Klapper stated in his reknowned *The Effects of Mass Communication* that the chief effect of the mass media was to reinforce existing perceptions. Twenty years worth of subsequent studies have not dented this conclusion; when George Comstock in his *Television and Human Behavior* reviewed the academic literature to date on the subject of television and voting, he summarized, "Crystallization and reinforcement, but not change and conversion, are the effects typically corroborated by scientific evidence." To the limited extent that viewers surrender attention to stories about political campaigns and candidates, they appear to use that material not to arrive at well-considered decisions but to prop up previous choices.

A fascinating study of television's role in the 1972 presidential election was executed by political scientists Thomas Patterson and Robert McClure, and published as *The Unseeing Eye: The Myth of Television Power in National Elections.* They conducted 2,000 hour-long interviews with voters before, during, and after the campaign. What they discovered upon analyzing their data was that people who had watched the political coverage regularly were no better informed on the issues than others who had not bothered to view. The added exposure to televised political information had left no trace. Apparently voters knew all they cared to know before the telecasts came on, and simply shed the political material. Perhaps this nil effect occurred because information is not what Americans seek from the medium.

Since entertainment is largely what people turn to television for, there is always the danger that viewers will slip further down Wiebe's scale and, as well as converting the news into *maintenance* messages, will also convert it into *restorative,* escapist fare. Mark Robert Levy observed on the basis of his interviews with news viewers, "Many people also find that television news entertains while it informs and reassures. Like situation comedies and detective shoot-'em-ups, the newscasts temporarily release some members of the audience from the pressing cares of daily existence." Looking at campaign coverage as entertainment, viewers may be more likely to focus on the horse-race aspects than on the significant issues. When Thomas Patterson went on to study the 1976 election much as he had the 1972 one, he first divided broadcast news items into two categories: game stories (on winning and losing, strategy, logistics, appearances, hoopla) and substance stories (issues, policies, traits, records, endorsements). Viewers were then asked by Patterson what election news stories they could recall. Not surprisingly, the items dredged up were approximately three times more likely to be game stories than substance stories. Although the newscasts also exhibit a bias toward game stories, the items which stuck in voters' minds exceeded that frequency. People do prefer the excitement of the race to the sober matters of policies and issues.

The tendency to take political news as entertainment has prompted many concerned social analysts to worry that the candidates who get elected are those who best suit the public's fantasy needs, rather than the requirements for government service. Ever since Richard Nixon's bearded, jowly showing in his 1960 debate with John Kennedy, all candidates have preened themselves for television appearances as if they were movie stars. Nixon himself, according to journalist Joe McGinnis in *The Selling of the President,* was able to recoup and to get himself so well decked out for the media in 1968 that he managed to lure the electorate's votes. (The evidence on this point is less certain than McGinnis suggests; Nixon started off with a 15 point lead over Hubert Humphrey in the opinion polls and, after spending $20 million at what McGinnis called "adroit manipulation and use of television," was by the first week of November barely one point in front.) Given the enormity of presidential responsibility in this precarious age, and the fact that campaigns are now conducted via a medium of entertainment, the resulting question remains a discomforting one: is the most likely winner the candidate who most resembles a matinee idol? Are we getting behind fantasy heroes when we should be getting behind real-world leaders? The election of Ronald Reagan in 1980 did little to silence this concern.

Inadequate time with the news, shallow attention when it's on, misinterpretation of the few stories seen, the conversion of important political coverage into entertainment—it is a poor record that Americans have with televised information. However, there is some evidence of recent improvement, for the size and intensity of the news audience has been increasing slightly, and portends to continue doing so. Since news-viewing is an avocation of the elderly, the ratings for informational programming are highly sensitive to demographic shifts in the median age. As the baby-boom generation matures and as the dropping birth-rate limits the downward pull of recent arrivals, the average age is slowly inching upward, and so is the receptiveness for information. This expansion is somewhat stimulated every time world events turn worrisome, as when American hostages were held in Iran.

The great popularity of *60 Minutes* is an indicator of the expanded tolerance for information. Much of the program's success results from its ideal placement within the weekly schedule, for the television audience on Sunday evening is larger than at any other time. Since Sundays are the one day when a person's normal number of *directive* messages is low, more information can be stomached then. Also, if an American family is ever going to view together, it will be Sunday evening, and a show like *60 Minutes* can be viewed by all without misgivings. An additional reason for the program's achievement is that it does not stay exclusively with *directive* content; some of its exposés perform a *maintenance* function by confirming viewers' sense of how the world really works. Plus, an abrasive *60 Minutes* reporter can be the vehicle for viewers' assertive, *restorative* needs.

Further signs of the expansion of news viewing come in the form of viable cable news services such as the UPI Financial News and Cable News Network. Every season that Ted Turner's CNN endures is added evidence that Americans will be watching more news as time passes. Responding to these early challengers for the future news audience, the networks are honing their operations by using vacant late-night periods for additional newscasts.

Nevertheless, news and information continue to make up the smallest portion of television-viewers' time with the medium. The networks try to interest people in more, and according to FCC figures devote about 15 percent of programming to it—but Americans resist. At most, 10 percent of total viewing time is spent with programs which attempt to add to the audience's store of information. It is the real-life experience of most adults that in order to lead successful lives, very few after-hours *directive* messages are usually needed. The majority intuit that their well-being is better served by fantasy shows. At the end of the working day most Americans have had their fill and are looking for ways to ease their burdens, not add to them; the further imposition of informational programming is not welcome. This is why the audience watching CBS for the disclosure of J.R. Ewing's assailant in November 1980 was 10 million people larger than the total audience for all three network broadcasts on the unpredictable 1980 Presidential election two weeks earlier.

Information Sent

Media Snobs, focusing on the activities of news broadcasters rather than on the appetites of news viewers, have laid down a barrage of criticism. They are convinced that it is the senders and not the receivers of information who are culpable for shortcomings in the system. First off, they point out, only a scant few stories can be aired in the 22½ minutes of broadcast time. And the stories that do make it on to the program are thought to be chosen more for their zesty film footage than for their intrinsic merit. Erik Barnouw, the sometimes scathing historian of broadcasting, writes, "The camera, as arbiter of news value, had introduced a drastic curtailment of the scope of news. The notion that a picture was worth a thousand words meant, in practice, that footage of Atlantic City beauty winners, shot at some expense, was considered more valuable than a thousand words from Eric Sevareid on the mounting tensions in Southeast Asia." The demand for exciting shots supposedly led to the exchange cited by Edwin Diamond in *The Tin Kazoo: Television, Politics and the News* where two news executives were comparing their coverage of a fire at a Roman Catholic orphanage to the coverage on another channel; the first executive complained, "Their flames are higher than ours," but the second countered, "Yes, but our nun is crying harder."

In the eyes of many Snobs, it's a network's drive after high audience numbers which means that the evening newscasts can never be more than shallow and sensationalistic. Frank Mankiewicz, counselor to the Kennedys before becoming a public broadcasting executive, claims in his book *Remote Control: Television and the Manipulation of American Life* that television news is governed by "an overriding law—The Trivial Will Always Drive Out the Serious."

Other critics perceive television news to be corrupted not so much by a headlong search after the most massive viewership as by its treacherous promotion of mammoth, repressive institutions in American life. An inferred kinship among Big Business, Big Government, and Big Television was one of the things that made Jerry Mander frantic about the medium in his *Four Arguments for the Elimination of Television*. When they can't influence the news directly, these established institutions are felt to shape it in a sinister, roundabout way through the machinations of their public relations specialists who are able to arrange for the staged and filmable happenings that historian Daniel Boorstin several years ago labeled "pseudo-events."

There can be no denying that the time allotted for network newscasts is brief. From 1963 until ABC experimented with its late hour *Nightline* program in 1980, the evening news broadcasts exceeded one half hour on just the rarest of occasions. Only about a dozen stories and a limited number of words can be crowded into those thirty minutes among the commercial breaks; both Snobs and promoters (Walter Cronkite is one) have pointed out that the word count on an average program amounts to less than half the front page copy of a newspaper.

From the beginning the networks have yearned to expand their newscasts. Prestige has been one reason: network personnel are not insulated from the Snobbish conviction that informational programming is more laudable than fantasy. They would rise in the estimation of everyone concerned, themselves included, if they could offer more public affairs telecasts and less socks-and-yocks. The ghost of Edward R. Murrow haunts network corridors and inspires awe and ambition.

Moral encouragement would additionally come from the Federal Communications Commission, which has long pressured the networks to transmit as much news and information as conditions permit. Committed by its founding legislation to democracy's ideal of an informed public, the FCC always examines a station's record of informational programming when its license comes up for renewal. Networks would be glad to acquiesce to this concern and secure the FCC's blessing.

The crude economics of news broadcasting also favors expansion. News is far cheaper to produce than entertainment—it can be put together at about half the per-minute cost of a prime-time fantasy. And once the news division has been set up, and the headquarters organized and the camera crews established and the correspondents on the job, then the expenses for filling increased air time drop proportionally lower and lower. Enlarging the evening newscasts to an hour would entail only a small increase in production costs, not a doubling.

Added incentives for the expansion of television news appeared in the late 1970s as the future of network broadcasting began to look murky. Executives at ABC, CBS, and NBC came slowly to realize it was within the realm of possibility that fantasy programming could defect to other modes of transmission—cable, cassettes, and satellite broadcasts among them. After all, the networks did not own the Hollywood production houses, and conceivably another carrier could outbid them for the serials, or worse yet, unpredictable regulatory decisions could rob them of that type of programming. And while the networks did produce sports broadcasts themselves, there was nothing to say that the leagues couldn't sign up with the competing, burgeoning cable systems if they chose to. The one content area the networks did control from start to finish was news. The more they could bolster this sort of broadcasting, the better position they would be in to face uncertain times.

Yet two countervailing forces have been muscular enough to hold the network newscasts down to half an hour. The lesser of these is the combined resistance of the local stations. Although everyone knows it is the national networks which have control over the finances and programming of the broadcasting system, from a legal perspective power is vested in the local stations. Technically it is the local stations, not the networks, which are regulated by the FCC and have standing in jurisprudence. In recognition of the lofty goal of community service, local stations are encouraged in communications law to act independently. Where they often choose to exert their prerogatives is at the moments they feel the networks are impinging on "their time"—in this case, the slots just before prime-time broadcasting begins in the evening, when local stations can transmit their own programming and collect all the advertising revenues for themselves. Sometimes the stations run homegrown programs or syndicated shows for the after-school and after-work audience, but often it is the local news, which costs comparatively little to produce and is generously profitable for the local owners. Station managements were irritated at losing fifteen minutes of this valuable time in 1963 when the networks twisted their arms and got them to carry a half-hour national newscast; they don't want to give up any more to the networks.

NBC came close to succeeding in 1981 when its affiliate board accepted a network proposal for an hour news program and recommended it to its constituency, the local stations; the 215 affiliates, however, balked at the idea. One local executive said, "Either we had to cut our own local news broadcasts down or move them to an earlier, less desirable time. We really don't see why the networks couldn't expand their news into their own prime time."

Why don't the networks expand the national news into one of the hours traditionally allocated to their own prime-time broadcasts? The reason why not, as is the case with much about broadcasting, is economic. Information simply attracts a smaller number of viewers than entertainment; thus advertising dollars would depart; thus income would shrink. A prime-time daily news hour would be a costly if not ruinous maneuver for a network.

The second and binding constraint on the expansion of television news, therefore, is viewer uninterest. As much as the networks want to, they cannot get Americans to view much informational programming. This explains why in 1979 the networks, despite all the incentives otherwise, had cause to put only 8.6 percent of their total payments for programming into the production of news and documentaries.

Restricted to a half hour, and to a smallish and fidgety audience, network news personnel still work hard to create as much viewer involvement as possible. To do this they have to turn to the medium's greatest natural advantage—the capacity to transmit visual images. In scanning their environment human beings, unlike moles or ants, are creatures highly predisposed to what vision brings in; these signals are the ones that penetrate minds the easiest and convey the most. As much as Media Snobs may disdain it, television's single, sterling compensation for the brevity of the newscasts is the accompanying film footage. Herbert Gans remarked in the course of his investigation into American news systems, "Television journalists see themselves as providing a headline service, which is meant to supplement the newspapers; but their main purpose and competitive weapon is the offer of 'immediacy,' bringing the viewer 'into' or near important and interesting events through the use of film." The demand for "good pictures" is the battle cry of every successful television news producer.

Getting good pictures is not easy work. Television news may be thought of as modern, snazzy, crackling reporting, but in truth it is less mobile and supple than old-fashioned newspaper journalism. By the time the camera man and sound man are set up, and the correspondent has his words ready, the story could be over. The line from legendary CBS news producer Fred Friendly that "Reporting the news on television is like writing with a one-ton pencil" could be taken as a comment on the influence of network news, but it is also a reflection on the technical difficulties of gathering that news.

It's not just the problem of capturing images on film or videotape which hobbles television news gathering. The footage has to be relayed back to the news studios in New York City where it will go through a lengthy editing process until the very few frames which contain the essence of a story have been pieced together. The people working on it scan the footage for the most dramatic shots, the 5 percent that will finally get on the air. Action footage is sought to balance off the large number of interviews, or "talking heads," that the newscasts are forced to carry—forced because an interview with an eyewitness or participant, or the stand-up remarks of a political figure, is often as close as video news gatherers can get to a story.

During the editing process the dramatic aspects of a news account are going to be enhanced, as a concession to the audience's known predilection for well-formed, diverting stories. Film editors are looking not only for striking shots but also for the material that will lend the item a dramatic structure. William A. Henry III, onetime television editor of the *Boston Globe,* tartly observed that among the unspoken principles of news production was this one: "Every story ought to have a dramatic unity, a clear line of conflict, with definable antagonists (reduced ideally to Homeric epithets), and a tangible prize at stake."

The classic examination of television coverage being shaped for its dramatic quality is a study done by Kurt Lang and Gladys Engel Lang of the 1951 MacArthur Day festivities in Chicago. Shortly after President Truman relieved him of his duties, General Douglas MacArthur came to Chicago to be honored in a parade and ceremonies. The Langs arranged to have thirty-one observers stationed along the parade route and at the speech site. Other observers simply watched the television coverage of the day. Later the two sets of observations were compared.

In the television version the General was depicted as a celebrated hero, a true American standard-bearer. Crowds were seen to welcome him enthusiastically, and to respond excitedly to his speech. In keeping with Henry's principle, the coverage had a story line with a beginning, middle, and end—from MacArthur's arrival at the airport, through his parade, to his speech and the concluding ceremonies. The antagonists were clear—MacArthur vs. Truman, the military vs. the executive branch, bellicose assertion in Korea vs. diplomatic restraint. And the prize, in the form of the laurels of public acclamation, was clear.

But the thirty-one eyewitnesses experienced little of this drama. The crowds which looked so dense on the telecasts were thought by the observers to be sparse-to-moderate. And to be lacking in exuberance also; the people along the parade route were judged to be there more out of curiosity than because of any heart-felt desire to honor MacArthur. The general, in fact, was scarcely visible to the eyewitnesses.

Television, the Langs concluded, had not conveyed the reality of MacArthur Day but had structured it in the interest of a dramatic presentation, one the at-home audience would find captivating.

The dramatizing of reality by video news-gatherers reaches a crescendo every four years during the televising of the national nominating conventions. What typically used to be a sputtering, meandering convocation, long on ho-hum tedium and short on punchiness, has become in the television era a prime-time contest of conflicting candidates and their partisans, going at each other with the tactics and exertion of football teams—that is, if the television reporters can make it seem so. Floor correspondents know that if they do not present conflict to the audience, viewers will defect in droves to better entertainment elsewhere on their dials, as they did when the 1976 Democratic convention was overwhelmed in the ratings by the All-Star baseball game. Drama is what viewers want to see at the conventions, and drama is what the networks angle to provide. In telephone interviews with 400 viewers of the 1976 conventions, it was learned that viewers had been far more interested in the conflicts than in the issues, and could recall stories about the convention's dramas over stories about the issues by a ratio of 4 to 1. The political scientist who conducted this study commented that "viewers prefer soft information, features, and entertainment from their politics. This is not lost on the broadcast industry, who appear to be giving the audience just what they want."

In trying to meet the audience's appetite for good pictures and dramatic stories, the way network news producers daily confront the problem of obtaining usable film or tape is through very careful planning. As much as people may want to conceive of news as something that breaks and then is speedily presented to them, the networks cannot operate in that fashion. It may sound ironic, or even deceitful, but an evening news show is planned a ways in advance—days, if possible. When John Chancellor said, "In this job, you find there are basically two kinds of days: days when you have a choice and days when you don't," he was being truthful but perhaps misleading. The days when producers have no choice and a newsworthy, unexpected event is thrust upon them—as the day that John Kennedy died—are so rare as to be almost nonexistent. In Edward Jay Epstein's masterful treatment of television news, *News from Nowhere,* he relates that of the news stories he analyzed only

2 percent were not anticipated beforehand. Almost all news days are characterized by advanced decisions and choices. Av Westin of ABC News was opposed to "waiting for news to happen in order to scramble after it. Anticipating events is most important."

Thus television news is indeed susceptible to "pseudo-events," just as Media Snobs charge. For the correspondent and crew to be at the same place at the same time as the newsworthy person, a great deal of advance information and prearrangement is needed. Press conferences and other pseudo-events are a means to vastly increase the efficiency of news gathering: a crew that wanders around randomly waiting for news to break would end up with next to nothing, while another crew tightly scheduled into one pseudo-event after another would bring back reels of usable film.

Do pseudo-events operate to the advantage of the large organizations and institutions that lurk behind them? Or to broaden this question, is television news the minion of Big Business and Big Government? Let's take these one at a time. The evening newscasts usually carry few stories about business either pro or con; a count conducted in January 1980 found that only 12 percent of all items were concerned with business. Thirty-one stories were analyzed, and 18 of them were found to be neutral toward business, eight unfavorable, and five favorable. No pro-business stance was uncovered here.

In fact, at that time the relationship between Big Business and Big Television could hardly have looked less chummy if one singled out the oil companies' dismay with the networks. The acrimony began in the fall of 1979 when newscasters reported on large third-quarter profit jumps for major oil firms. Mobil, Shell, and Texaco felt they were being picked on and retorted through press releases and newspaper advertisements. A Mobil spokesman thought he knew what the networks were up to: "They report bad news in sensational ways to win ratings." But according to Herbert Gans, the reason the story had been carried was simple enough: while American consumers were suffering because of rising gasoline prices, the oil companies did not seem to be sharing in that pain. "In fact," Gans wrote, "the story could have been left out only if the newscasters were prepared to lose their most precious resource: their credibility with viewers." Gans's point is that it is the interests of viewers, not Big Business, which are served by television news.

The situation is not dissimilar with regard to Big Government and its most conspicuous symbol, the President. It's curious that even though the President is the chief source of news in the United States, and the news system very much needs the stories he provides, still their relationship is hardly close and friendly. The institutions of the news and the Presidency are locked into a ceaseless tussle, with the President often feeling misrepresented and defensive and the news gatherers feeling toyed with and misused. In other nations the leadership and the press suffer no such rifts since the government's interpretations are relayed as the official news. But in this country the newscasts are not the conduit of Washington's version of things. Just the reverse: the press is the public's representative in court, present to ensure on behalf of its constituency that everything is going as it should be. When the news gatherers are nosing around, it is generally in the interest of the public and serves as a check on political power.

If the relationship between the news media and the President becomes too strained, the President usually grows distant and petulant; there isn't much else he can do. But Richard Nixon, operating through his vice-president Spiro Agnew, ventured a different, novel reaction—a counterattack. Nixon had been stung by the networks' resistance to his progress report on the Vietnam War delivered November 2, 1969; news commentators had immediately afterward voiced skepticism about the President's version of how well things were proceeding toward a conclusion. The task of retaliation was delegated to the willing Agnew, who took the occasion of an otherwise unremarkable meeting of the Midwest Regional Republican Committee in Des Moines on November 13, 1969, to assail television news.

The Nixon Administration had challenged the networks to carry Agnew's speech, and for a variety of reasons—some praiseworthy, some cowardly—all three did. In the minds of the total American audience Agnew attempted to plant the notion that network correspondents had an elitist view of national affairs which was detrimental to the country's well-being. "Are we demanding enough of our television news presentations?" he asked at the beginning of his speech. The newscasters and producers—"this little group of men" he called them—"not only enjoy a right of instant rebuttal to every Presidential address but, more importantly, wield a free hand in selecting, presenting, and interpreting the great issues in our nation." These newsmen, he went on to say, "live and work in the geographical and intellectual confines of Washington, D.C., or New York City." At the climax of his speech Agnew insisted, "The views of the majority of this fraternity do not—and I repeat, not—

represent the views of America." The result of their "endless pursuit of controversy" is that "bad news drives out good news." Agnew urged his viewers to contact their television stations and networks and complain.

The speech was a serious challenge to the role and viewpoints of network newsmen. It is estimated that 150,000 people wrote or called the networks and that two-thirds of them supported Agnew's position. But the loss of faith, if it ever existed, was short-lived. Within a month an ABC poll revealed that 60 percent of the public thought the networks were being strictly fair with the White House. Bower's 1970 survey also reported that the majority of Americans believed television news was politically unbiased. Another study done a year later put it this way: about one-half of the public detected no bias, while one-quarter believed television news was inclined toward the Nixon administration, and the final quarter perceived bias against. Agnew's speech, together with other administrative tactics, prompted Walter Cronkite in 1971 to refer to "a grand conspiracy to destroy the credibility of the press," but the public appears not to have been taken in. The news, they sensed, had their interests at heart.

To serve viewers, the news must consist of good and bad alike. A researcher intrigued by Agnew's assertion that "bad news drives out good news" actually set out soon thereafter to have fifteen days worth of news stories from the three networks counted and categorized. Stories put in the "bad news" category were about 1) the Vietnam War, 2) international tension, 3) social conflict, 4) crime, 5) accidents and disasters, and 6) other miscellaneous bad news. Of the 820 news items sorted by trained researchers, just 34 percent could be classified as bad news.

The news content the audience wants comes in response to an unarticulated question about the continuities of life and can report either on those continuities being threatened, in which case the news is bad, or on the continuities being enhanced, in which case the news is good. If surveillance results in bad news, the audience is forewarned; if good, the audience is reassured. People tend to remember the bad news because it calls for some sort of response, but good news is just as crucial in understanding the state of the world.

It is only reasonable that television news act in behalf of its audience, and not in behalf of Big Business or Big Government. The close identification of viewers with newscasts is what networks must aim for if they are going to have the maximum number of people to deliver to advertisers. But does the pursuit of viewers' favoritism and the highest possible ratings pervert the integrity of network news? Does it lead newscasters to dwell overlong on the splashy and inconsequential? Most critics would say yes.

Even the loyalest supporters of network news have to concede that, like all television programming, it too is engaged in heated competition for viewers' attention. Walter Cronkite declared during an interview, "Let me say right here that I am not one who decries the ratings. Those among us in the news end of the broadcasting business who do are simply naive. Of *course* ratings are important, and no one—newsmen, program managers, salesmen, or general manager—need hang his head in shame because this is the fact."

While television news' pursuit of large audiences and advertising revenues is thought to be suspect, no one seems as bothered when the same thing happens in print journalism. Shades of Media Snobbery. Don Hewitt, the producer of *60 Minutes,* pointed out the double standard: "Why is 'circulation' a clean word and 'rating' a dirty word? Newspaper columnists talk about the decibel level of commercials versus the rest of the program. Why did the print ad for the used car dealer in Louisville run in bigger type than the headline 'John F. Kennedy Assassinated in Dallas'? Why do I begin a story on page one and have to look for the rest of it somewhere between an ad for a special at the A&P and an ad for a new Sears refrigerator?"

In any case, the way high ratings are best won for national newscasts, it seems, is through the sober, accurate recounting of newsworthy events and personages. Evidence exists that television reporting is even more accurate than print. Researcher Michael Singletary checked the precision of both television news stories and newspaper accounts by having the items sent to the people mentioned in them, and asking them to determine how truthful the report was. About two-thirds of the television stories were judged to be "entirely accurate" by their subjects, while only one-half of the newspaper ones were. Viewers' faith in the credibility of television news does not seem to be misplaced.

In network newscasts it is only infrequently that the sensationalistic will displace the substantive. David Brinkley responded to charges that too much coverage was given to the 1977 siege of the Hanafi Muslims in Washington by saying, "It possibly is true that public attention encourages this kind of violence and we are very conscious of that. But on the other hand, when this kind of terrorism appears in a country once largely free of it, I think the American people need to know it." Brinkley was indicating that there are times when the two tasks of the news—to alert and to calm—come into conflict, and that at those moments the filmed action may be too stirring for some viewers. More often, however, the news will stray in the other direction, in the intent of reassurance, and try to soothe public fears; this is what initially occurred during the Three Mile Island nuclear incident in 1979. Producers know that if their stories are inappropriately strong, they risk damaging their reputation and losing more viewers than they would gain. NBC's "Policies and Procedures" manual for newsmen warns them away from these dangers: "News may never be presented in a manner which would create public alarm or panic."

In response to the Snobbish accusation that news broadcasts are pitched too low, an interesting counterproposition has come from James David Barber, professor of political science at Duke University and author of *Presidential Politics.* Barber first reminds his readers that the majority of Americans, while saying that television is their best-liked news source, don't bother to watch it regularly. Only about one person in 50 views the network news every evening, Barber recounts, and of those only one in a hundred says he gives it his "full attention." Barber then goes on to say boldly, "The trouble with television news is that it's too good—too intellectual, too balanced. It passes right over the heads of the great 'lower' half of the American electorate who need it most. If those who would reform television stopped thinking in terms of turning the network news into the *Encyclopedia Britannica* or the *New York Times,* it could realize its enormous, unexploited potential for reaching and enlightening voters who now do not know what it is talking about." Arguing that people who cannot understand the vocabulary will turn off the newscast, Barber found these terms on an ordinary *CBS Evening News* program in 1979:

> allocation formula
> collusion
> surcharges
> most-favored-nation trade status
> tariff concessions
> wage-price guidelines
> honoraria
> cottage industry
> trade credits
> litigation

Somewhere in between Barber's perception and Snobs' perceptions is probably where the news is really targeted, to strike the great middle ground of American taste, knowledge, and interest. Television news has arrived at a balanced approach, neither too cerebral nor too gaudy, by which the greatest amount of information is conveyed to the greatest number of people. If the news were more ponderous, the audience would be smaller still, and if it were more flamboyant, its long-term responsibilities would not be fulfilled. As things now stand, what tolerance for information there is on the part of the American public is being filled if not widened.

In addition to its relative lack of bias and its wealth of good pictures, the most positive feature of television news is its ability to expand its coverage and broadcasts on those infrequent occasions when crises loom or important national events transpire. Suddenly video news gathering efforts will begin to concentrate on a particular story, as the audience's concern mounts. It is then we realize that television news is one of our greatest resources in reserve, capable of deep and critical influence. At these times, for these matters—some of them brutally abrupt, like Kennedy's death, others torturously prolonged, like the Vietnam War— television reports penetrate viewers' minds to affect their perceptions and judgments, to trigger the processes of public opinion and mobilization that eventually produce resolution. These are the stories that refuse to go away until surveillance can finally lead to reassurance. ABC's Bob Shanks asserts, "I credit television with stimulating the civil rights movements of blacks and women, . . . shortening America's participation in the Vietnam War, . . . and hastening the fall of Senator Joseph McCarthy and President Richard Nixon—without shredding the social fabric."

The fall of Richard Nixon—that was an event profoundly influenced by television journalism. "Without doubt the major news story of the decade is the constellation of events which has come to be known as Watergate," remarks David Altheide, another of the group of social scientists (which includes Herbert Gans, Edward Jay Epstein, and Gaye Tuchman) who have taken broadcast news as their object of investigation. Ostensibly the Watergate story was that of minor political tomfoolery—Republican-hired thugs caught in the act of skulking around Democratic Party headquarters in the Watergate building complex after hours on June 17, 1972. But those who peg this chapter of American history to that incident will remain as mystified as Nixon himself was about the subsequent chain of events. The real story was the reactions and defenses of the President. Witnessing him through the media, the public came to feel he was acting as if he had something to hide. In time it became clear that he did.

What Nixon needed to conceal was an odious and potentially treacherous personality which the electorate began to feel was not suitable for the nation's highest office. Reelected in 1972 by an enormous plurality, Nixon had been the public's choice to stave off a perceived danger on the left. But once the McGovern threat had been silenced, questions came to the fore regarding Nixon and his associates. Were they the sort of people to entrust the nation's course to? How principled were they, or weren't they?

The investigation into the President's affairs began through the initiative of two journalists and appeared in one newspaper; as such it had a minute effect at most upon public consciousness. Not until the story finally surfaced on the network news were the mechanisms of public opinion formation set into motion. The 37 days of live Watergate hearings in the summer of 1973 served only to stir up, and not to squelch, the apprehensions of the American people. So attentive was the audience that the displacement of daytime fantasies was not met with the outcry it might have been at other times. Viewers were learning about a President who appeared more deceitful than other Presidents in their experience; each previous President had suffered his embarrassments, but none reacted as defensively as Nixon, nor was surrounded by a staff as seemingly sly and devious. When the tape recordings of Oval Office conversations were released by the President in a televised speech the following year, the public did not like what it was seeing and hearing. Correspondent Sander Vanocur sums up, "Without television's coverage of Senate Watergate hearings and House Judiciary Committee hearings the following summer, the American people would never have changed their attitudes so drastically that a reluctant Congress could safely proceed toward the formation of articles of impeachment."

The downfall of Richard Nixon was due to the nation's need for certain qualities in its leaders and his personal shortcomings in that regard. What television news did was to put the President and the public in touch with each other through "good pictures." Reassurance finally came with what some feel are the best, if not saddest, shots of all: Nixon pausing before entering his helicopter on August 9, 1974, gamely waving goodbye. That was one of the few evenings that the news, usually unappreciated, became important to Americans.

Commercials

For some few members of the audience the most informative content on an evening news show may not be what the correspondents report but what the commercials display. This is more likely to be true for older than for younger viewers, as the products advertised among the news stories run toward denture cleansers and laxatives, arthritis medications and digestive aids, sleeping potions and decaffeinated coffee. Manufacturers of these items knowing that the median age of news viewers is high, make use of relatively inexpensive spots on pre-prime-time news programs to lay out their wares. Nostrums are not all that are advertised, though: other notices try to sell airline tickets, tonic water, stockbroker services, traveler's checks, and so forth. Somewhere in the massive audience are people for whom these messages will seem to be especially composed. Whether their dental plates are loosening or they are about to jaunt around the world, or both, the commercial shows them something they may want to know about.

This communication between producers and consumers has to be understood in the light of the considerable affluence of Americans. When compared to the purchasing power of people in most other countries of the world, the wealth of individuals in the United States remains high. Americans may complain about economic fluctuations and the bite of taxes, and feel that their expectations for gain are not always met, but through thick and thin, in spite of inflation, the *Wall Street Journal* reports that our personal finances have continued to improve throughout the television era. Disposable income (that is, what a person makes over and above what has to be allocated to basic necessities) was about $2,000 per capita in 1960, over $3,000 in 1970, and topped $5,000 in 1975, in constant dollars. Simply put, Americans are people with money to spend. The question is how to part with it most advantageously; this is where the words from the sponsors come in.

Commercials can convey various kinds of information relevant to the act of purchasing. Television specializes in the promotion of new products, so the very fact that something is being offered in the marketplace which wasn't there before can be informative. Whether old or new, the product is usually shown, along with its packaging, so that Americans can get an image of it. Frequently its intended uses are illustrated— the bowl of cereal is being eaten, the recreational vehicle is bounding over hillsides, the brassiere is on a mannequin. Distinctive features will be mentioned, even if it's only an unintelligible but extra ingredient, or a particular color. Of course there's a great deal about the product that is *not* communicated: viewers normally don't learn its price, or its shortcomings, or how it compares to all of its competitors. But one way to estimate the worth of the skimpy product information the audience typically does get is to recognize that, as little as there is, it comes effortlessly. People don't have the time or energy to search out all they need to know to make the wisest purchase each time. The sheer convenience of the televised information is one thing to its credit.

Besides the product information, much of an average commercial plays to the psychological needs of the consumer which the product may help to satisfy. Viewers learn of possible ways to assuage deep-lying needs and drives. Need love, for example? Buy food for a pet. The more affluent Americans have become, the more we have wanted to indulge ourselves by bathing in the devotion of a dog or cat. Pet food producers gleefully watched their sales increase threefold from the '60s to the '70s. To keep their names before consumers and to avoid being outdone by their competitors, and perhaps to jack up the overall demand faster than was happening naturally, the makers of dry and moist, bagged and canned nourishment for small animals rushed into television advertising. "Never in the history of television, with its news coverage emphasizing wars, invasions, fires, bombs, political chicanery, and its dramatic programs emphasizing shoot-em-ups and missions impossible, has love been promoted so assiduously as in the pet-food commercials. The whole panorama of pet-food advertising is awash with sentiment, tailwagging enthusiastic woofing, contented purring; it oozes domestic coziness," observes *New Yorker* writer Thomas Whiteside. Talking dogs and tap-dancing cats are presented as the cute humanoids that Americans can cherish, to be loved in return. Viewers want love and can pay for it; the commercials show how to achieve it, and suggest which brands will make the pet most perky and adoring.

Beyond product information and information regarding possible gratifications, viewers are led to make inferences about the availability of whatever's being advertised. It is implied in the message that a person can go shopping with some assurance the product could be found.

In the view of one economist, however, the most telling information provided by a commercial frequently has nothing to do with product description, or suggestions regarding a linkup with needs, or implications of availability. Instead it is the very fact that the manufacturer bothers to advertise at all. Professor Phillip Nelson of the State University of New York at Binghamton states that on the whole advertised products are better buys than unadvertised products. "The better brands have more incentive to advertise than the poorer brands. It is the market power of the consumer to repeat purchase the brands he likes that makes the difference. Those brands that get a lot of repeat purchases find it more profitable to advertise than brands that will not get repeat purchases. Simply put, it pays to advertise winners rather than losers. In consequence, the amount of advertising gives consumers a clue as to which brands are winners and which brands are losers."

While viewers can pick up various kinds of direct and inferential information from any commercial, the actualities are that most of the time they don't. Advertising messages are mightily sent, but little received. The 30-second spots are the least viewed and absorbed of all programming. When researcher Robert Bechtel made his videotapes of people viewing television, he learned that commercials were watched only 54 percent of the time they were on (which was 1 percent less than the next least-watched content—news). The recall of a particular commercial seen the evening before will typically drop below 20 percent of the audience.

There's all good reason for commercial messages to be deflected. Their purpose is to get viewers to make purchases of a product that is in all likelihood unwanted, and if desired, may be prohibitively expensive. It is in the nature of the American economy that more goods will be offered than will be purchased, so the consumer is often in the wearisome position of being besieged with commercial enticements. The few Russians who manage to emigrate to this country find the shower of advertisements to be a wonder beyond belief, and for days will study their new television sets to see all that is being offered in the marketplace. But soon enough they too become inured. They learn to do what all good television-watchers do, which is to ignore most of the 30-second entreaties.

Even though the commercials constitute what it costs viewers to get the fantasy entertainment they really want, still the public's attitudes toward them are more neutral than negative. Steiner's 1960 survey found few viewers who had any reaction at all to the commercials, either pro or con. When asked how they felt about the medium in general, only 7 percent thought to mention advertisements—1 percent to praise, 6 percent to criticize. Ten years later, in response to pointed questions about commercials, Bower's respondents confessed to conflicting attitudes: 43 percent agreed that commercials could be annoying, and 54 percent concurred that "some commercials are helpful and informative."

An extensive study of public opinion regarding all advertising was conducted in the mid-'60s by professors Raymond Bauer and Stephen Greyser of Harvard University's Graduate School of Business Administration. Their conclusion was, "By all the measures we have used, advertising does not occupy a central position in people's consciousness." Of the small fraction of television advertisements that managed to make an impression upon the subjects of their study, 27 percent were deemed annoying, 4 percent offensive, 38 percent enjoyable, and 31 percent informative.

Even more tolerance for television advertising is expressed by Americans when the interviewer places it within a context. A large majority of the public—80 percent in a 1979 Roper poll, and slightly less in the Steiner and Bower surveys—said they believed that commercials are a fair price to pay for the other programming they receive. And when television commercials are compared to other modes of advertising, as they were in a study commissioned by the Television Bureau of Advertising, they are judged by the public to be the most exciting, most authoritative, and most influential.

While attitudes toward commercials are not as negative as they might be, still the messages are not warmly welcomed. If the information is going to get through to viewers, it has to be sugarcoated. This is why commercials contain material relevant to the psychological needs of viewers in addition to material that serves to describe the product. The product information has to be wedded to what advertisers hope is an appeal to the viewer's psyche, to the underlying drives and motives of the individual consumer. In an on-target probe, Marshall McLuhan observed, "Gouging away at the surface of public sales resistance, the ad men are constantly breaking through into the *Alice in Wonderland* territory behind the looking glass which is the world of subrational impulses and appetites." Refundable travelers' checks are not just convenient currency; they become a hedge against catastrophe and evil. A vitamin and iron supplement does not just stave off anemia; it brings binding companionship along with a bit of nuzzling and embracing. A beer is not a mildly alcoholic drink; it's a reward, a good time with good pals.

The appeal that advertisers choose to associate their product with does not necessarily have to be upbeat or enticing. If headache remedies or sewer services are being sold, something abrasive and jangling may work, in that viewers will come to link an aggravating problem with the particular solution. After all, there is no known correlation between the effectiveness of a commercial and its popularity. Richard Pinkham of the Ted Bates advertising agency (the one responsible for the Anacin campaign) remarks knowledgeably, "Some of the most irritating commercials have been the most successful." Commonly, however, the associations are meant to be congenial and inviting. This product leads directly to something pleasant, is usually the message. Often the appeals are through scenes slightly displaced from the viewer's situation in time or place, invoking longings for greener pastures. Beaches, mountains, and other vacation sites sell many goods. New products are often marketed either by nostalgic appeals with front porches, general stores, rural scenes, and quaint, elderly spokespeople or by the opposite, with a modern, up-to-date life-style featured. Canned lemonade is sold the first way and most other soft drinks the second.

Whether the pitch is grating or ingratiating, it frequently comes in the form of a compressed playlet. Speedily, in 28 ticks of the clock, the action shifts from confounding problem to heavenly solution courtesy of the advertiser's product. These "instant passion plays," as someone has dubbed them, begin as the location is established and the vexation is emerging: the sink is stopped up, or the line is forming at the duplicating machine. Ten seconds into the commercial and the situation is out of control. Salvation: the pizza arrives, or the deodorant. We are briefly allowed to gaze upon the wondrous product and its packaging before returning to the scene where all is well and everyone is beaming. The closing shots are of the label and the product. Lesson summarily over. "What Japanese poets have done in the haiku, American admen are doing in TV commercials: concentrating an incredible amount of information and suggestions into capsule forms," says popular culture expert William Kuhns.

Like any drama, these playlets oblige viewers' need for relief from stress and strain. Other commercials are targeted at the need to be guided and directed. The message comes in an imperative form whose thrust is, "Trust this product. Buy it." Testimonials are the way this is often communicated. When a baseball hero praises a coffee maker, or a well-known actress touts a line of cosmetics, then the viewer may feel instructed to try the product. The same principle is at work when nonprofessionals ingenuously declare to the camera, "This product worked for me. It will work for you. Get it." And some of us do.

From a technical perspective, to impart the product information and the emotional appeal to the consumer with as much power and craft as possible is not an easy thing to do. Commercials are the product of more thought and skill than any other content on television. It costs as much as a quarter million dollars to produce a 30-second spot, which will then run between ten and a hundred times during a viewing season in time slots purchased for somewhere between $20,000 and $200,000 each, depending on the size of the audience. Almost every commercial is seen by more people than have been to every theatrical production put on in the United States since colonial times. For audiences of this size, at costs this high, advertisers want their messages to be perfectly wrought.

So well fashioned are commercials in conception and execution that sometimes they are thought to be better than anything else on the medium. It is not what they say, but how they say it, that lands them this appreciation. The favorites of 2,500 viewers queried in 1979 were the commercials for McDonald's ("You, you're the one"), Pepsi ("Have a Pepsi Day"), Anheuser-Busch ("You can call me Ray"), Miller Lite ("All you wanted in a beer and less"), and Purina Cat Chow ("Chow-chow-chow"). Ironically for advertisers, a percentage of the fans of a particular commercial are not potential customers at all; they are people who have already bought what is being advertised. "It's often been noted that those who most enjoy ads already own the products," explains Edmund Carpenter, Marshall McLuhan's fellow media visionary and the man who saw the correspondence between dream fantasy and television fantasy. "Ads increase participation and pleasure; they hold definite experiences. A product without advertising can be, for many people, a non-experience."

In their drive to get their message across, are advertisements fraudulent as well as beguiling? Is the scanty information also untrue? Without question there have been deceptive commercials on television. Zenith had to stop claiming "Every color TV Zenith builds is built right here in the U.S. by Americans" when it was discovered that 15 percent of the components were imported. In the tightly-contested pet-food market, Alpo led customers to think that its food was entirely meat by saying it contained only "meat-by-products, beef, and balanced nutrition"; "balanced nutrition," it turned out, was a copywriter's way of saying soybean flour. Anacin insisted it was a tension reliever despite having no evidence for it. Ford commercials featuring Bill Cosby stressed "tough cars" and "tough engines," but from 1974 to 1978 Ford manufactured almost two million defective small cars whose camshafts and rocker arms wore out. At Rosie's diner, the claim for Bounty paper towels was that they could support a cup and saucer when wet; so can any paper towel when gripped with the grain of the material. And the well-advertised STP, the gasoline additive, apparently accomplished nothing at all except to soothe customer's needs to do something quick and cheap for their automobiles.

Yet of the thousands upon thousands of commercials broadcast every year, only a tiny number of them are demonstrably untrue. The reason outright deception is so infrequent has nothing to do with the moral fiber of advertisers or their sense of the greater good; competitive pressures on them are severe enough to submerge any such qualms. Commercials have a hard time airing lies because they are scrutinized by a pack of steely-eyed watchdogs from the networks, the advertising industry, consumer groups, the federal government, and most fiercely, the competition. Additionally, there are the viewers themselves, who directly through complaints or indirectly through slackening purchases of a misrepresented product, exert control over deceitful commercials. Let's look at some of the forces at work which keep advertising information generally truthful.

The relationship between network "standards and practices" divisions and the advertising agencies, which used to be friendly, has shifted toward an adversary one, to the point that a network will turn down a commercial if they anticipate people will be misled by it. Broadcasters have learned that it is better business in the long run to keep the 30-second spot free of anything unprovable or truly deceptive, or anything that might make viewers more resistant and wary toward the medium than they already are. As a result, when common sense would lead one to think that the network would gladly take the money and usher all submitted commercials on to the airwaves, in truth advertisers have to run a gauntlet with their 30 seconds of high-powered entreatments. Commercials are first previewed by each network in their preliminary, storyboard form, and specific revisions will be directed. The soap powder cannot be treated as a panacea; the child's toy cannot be filmed at an angle which will make it look the size of an earth-mover. When the filmed commercial is delivered to the network, there must be affidavits on file swearing that nothing was rigged. The commercial is reviewed again to see if the original agreements have been kept and no disallowed innuendoes have crept in. If the actor in the patent medicine commercial looks too anguished, implying the remedy will cure more than it does, then the spot will be sent back for refilming.

From the advertiser's point of view, the process of review and revision would go much more smoothly if the three networks acted in concert. But they don't, and the agencies must oblige three slightly different sets of standards. The J. Walter Thompson Agency had trouble with a Kawasaki motorcycle commercial which showed a cyclist zipping across the countryside while rock music blared and rainbow-colored lights flashed. "One network said it looked like the guy was speeding, so we slowed him down," reports Howard Abrahams, a J. Walter Thompson lawyer. "Another said the lights were psychedelic, as though we were promoting the use of drugs. So we toned down the lights. The third said it was O.K. in the original form."

Once the commercial gets past the network censors and is broadcast, then the producers of competing products may have something to say. Heinz squealed on Campbell's trick of putting marbles in the bowl to lift the meat and vegetables to the top so the steaming soup looked chunkier. Competitors compelled American Express to reshoot dangerously effec- tive commercials featuring actor Karl Malden because viewers were getting the inference that the other issuers of traveler's checks would not make quick refunds in time of need. If such disputes can't be resolved in a gentlemanly fashion, they often end up with the National Advertising Division of the Council of Better Business Bureaus. A self-policing office of the advertising industry, the NAD will ask advertisers to withdraw commercials if claims can't be justified. Vacor rat killer promoted itself as "America's Number One Rat Killer" until the NAD reminded the manufacturer that d-Con was the best seller. If questioned claims can indeed be substantiated, then the NAD supports the advertiser; contrary to challenges, Freedent doesn't stick to dental work, and Krazy Glue does stick almost anything to almost anything else.

The court of last resort for complaints is the Federal Trade Commis- sion, which works sluggishly but with great finality, since its decisions are legally binding. It took 16 years of litigation before the FTC saw fit to make Carter's Little Liver Pills drop the word "liver." To stop Geritol from talking about tired blood was also the result of protracted delibera- tion. But the FTC can move more swiftly if the case calls for it. In 1977 it ordered the STP Company to pay a $500,000 penalty for false advertising, and an additional $200,000 for corrective advertising, because its gasoline additive had been misrepresented to viewers. To come back again to one of the bulwark advertisers on national news shows, Poly-Grip was rebuked by the FTC in 1978 soon after its commercials implied that people who used it could eat corn on the cob with abandon.

One of the ways advertisers try to fight back against the constraints, apparently with the collusion of broadcasters, is to have their messages broadcast at higher sound levels, Although agencies and stations deny it, viewers' ears tell them that the commercials are louder, pure and simple, than the surrounding programming. The commercials, sometimes battered and weakened in the review process, are still trying to penetrate the hubbub of the average living room.

Operating in an atmosphere of scrutiny and restriction, how effective can television commercials be? The answer is both *not very* and *quite a bit,* depending on how the situation is perceived. Only a very few viewers will be able to recall a commercial after seeing it. Not only don't commercials make an impression on us, but as strange as it may seem, no experimental evidence exists that they get us to buy anything. George Comstock, in his review of the scientific literature in *Television and Human Behavior,* says, "There are no publicly available studies which unambiguously relate changes in behavior to exposure to television advertising." In these ways television advertising looks to be unavailing.

And yet, on the other hand, it is clear that enough people retain the product information and make purchases that advertisers are willing to continue investing in television commercials. "There's no question television sells goods," blanketly states Robert Blackmore, NBC's vice-president for television network sales.

The secret in all this, the key to understanding this discrepancy, is the enormous size of the audience. A survey of a sample of viewers will discover the percentage who recall a commercial is minute, and experiments on a small group of subjects will find scarcely a trace of influence from advertising messages. However, when these tiny fractions are applied to an audience of over 30 million people, the numbers become large enough to account for many sales and further advertising. Horace Schwerin of Schwerin Research Corporation has said, "The fact is, all but about 10 percent of the money spent on television commercials is wasted. The amazing thing is that television pays out, running on an efficiency of 10 percent." Ten percent is probably an inflated figure—1 percent may be more like it. If, of all the people who see a repeated

commercial, 1 percent are moved to try the product, then the manufacturer, advertising agency, and network can be content.

A relationship between commercials and sales cannot be statistically
established because purchases result from a large number of factors,
including changing real-world cultural and economic conditions, the
posture of competitors, the retail availability of the product, the color of
the carton, and on and on. But new *perceptions* in response to a
television advertising campaign can be documented by survey researchers, and from time to time they tell of a clear-cut success. North-
west Mutual Life Insurance Company went from 34th to 3rd place in
public recognition of insurance firms through a $1 million, two-week
advertising blitz on the 1972 Olympics. "Hey, Goodrich doesn't have a
blimp," the spots declared, and F. B. Goodrich experienced a 222
percent increase in consumer awareness. Hallmark places almost all of
its advertising money on television, and now reaps the benefits of having
its name recognized by 98 percent of American women.

Needless to say, commercials don't always work, either because the
message is poorly wrought, or the audience simply doesn't care for the
product, or the product's story can't be told via television. Florists
Transworld Delivery learned early in television history that their service
couldn't be adequately described on the medium, and so for years stuck
to magazine advertisements instead. New York advertising man John
Warwick says, "There is no doubt that TV, when it works as a selling
medium, works beautifully. If there is a high emotional content in what
you are selling or in your way of selling it, then TV can be a medium of
high impact. But if your advertising is more utilitarian, with a lot of
detailed or technical information, TV may be a bad choice."

Nevertheless, most advertisers feel that if they are in the business of
selling products to individual consumers, television is the advertising
medium they must succeed in. They shovel billions of dollars annually
into the industry, and feel they have little reason to complain about
prices which are climbing at twice or three times the inflation rate. If they
wince at paying over $100,000 to rent one 30-second time slot on a
situation comedy, they don't do it in public, because they are aware that
other advertisers are standing in line, especially if the show is a popular
one. If anything, they feel lucky to have the chance to buy into the show,
since they know that the more popular a program is, the higher the recall
rate will be for the words from the sponsors.

So anxious are advertisers to crowd their way onto television that the medium is now second only to newspapers in advertising revenues. Newspapers can expand their pages to accommodate more advertisers, though, while television is stuck with the inelasticities of the clock. The result is pressure to increase the advertising spots at the expense of the fantasy programming, pressure met only by mounting audience resentment. During prime time the networks operate under a gentlemen's agreement to sell only 9½ minutes per hour, and local stations may add a minute or two. Under these conditions, the only way to handle more advertisers is to subdivide the advertising time into smaller units. This is what happened during the '70s, when the average spot changed from 60 seconds in length to 30. Since recall for a 30-second commercial is not one-half of what it would be for a 60-second one, but rather three-quarters, everyone involved from the broadcasting and advertising industries accepts this development.

On the average, each American is now exposed to over 1,000 spots weekly. It is an enormous menu. From the banquet of information—which is partial but not blatantly untruthful and certainly handy—most people choose according to individual tastes and resources the particular diet of goods and services that serves them well.

10
Television Is Good for Your Children

The Scare

When Senator John Pastore formally petitioned in 1969 for the now-famous investigation by the Surgeon General of television's effects, he wrote, "What is at stake is no less than our most valuable and trusted resource—the hearts and minds of our young people." He was exactly right: children *are* our most valuable and trusted resource. They are a nation's sole guarantee that a future exists and that life will endure. It's because they are so precious to us that concern about children's television runs so high.

Love is the parental emotion which endorses an infant and signals the child's high importance. If an adult is ever to cherish anything in life, it will be an offspring. Nothing else is likely to evoke such a sense of devotion and purpose. But hard on the heels of love comes the parents' duty to raise up a child so that eventually he will be able to step into the roles of the departing adults. It is strenuous work, and a test of parental affection, but the new member must be socialized. Drives must be channelled; skills must be taught; right has to be separated from wrong; the real world must be parcelled out and revealed to the child as the capacity to understand it grows. Through years of sanctions and encouragements parents mold the behavior of their young, bringing the child to the point where he can go it alone.

Socialization would proceed much more smoothly if the minds of children were different from what they are. Should children be more receptive and more attuned to reality, then the jobs of parents would be easier. But children are not: young brains are bubbling with vague and contrary motives which often hamper learning. When learning does occur, it can be a most uneven fashion—the child can learn what the parent least wants him to, or may have to learn again something that was previously mastered but forgotten. The starting point for learning is remarkably low; boys and girls don't understand things so simple that as adults we would never pause to consider them if children didn't indicate they were in need of these concepts: front and back, full and empty, present and gone. Most critically, children are unable to distinguish well between what is real and what is unreal. Where a flight of fancy stops and the real world begins can be unclear to them.

Television has come to deeply intrude into the difficult business of child-rearing. It has joined the other socializing forces which surround and shape young personalities, and in doing so it has probably forced the traditional agents of parents, other children, play, and print to give way a bit. A commanding presence in children's lives, the medium exposes them to things they would never see or hear otherwise. Without question it instructs children, if only because they seek instruction from every source. But their ability to discriminate proper instruction from improper, and reality from fantasy, is so circumscribed that the question of how television content affects young brains can only be a troubling one. All in all, the ambiguities of the situation mean that the alarms and barbs of Media Snobs will be listened to by anxious parents.

Marie Winn, author of *The Plug-In Drug,* concedes that it is probably all right for adults to watch television. They work, have grown-up responsibilities, and are able to handle whatever the medium dishes out. The entertainment and information content may even have some utility for big people. But as for children watching, that's another matter, Winn feels. She believes passionately that television shows are addictive and destructive for the new generation, that the medium catches hold of fresh minds and slowly, over thousands of hours of contact, blots them dry of energy and creativity. Winn talks about the *television savant,* the child who has seemingly learned jingles and facts from the programs, but who (like the *idiot savant* in psychological literature who could mentally solve elaborate mathematical problems yet was essentially empty-headed) really knows nothing of consequence and has not amassed the building blocks of further development.

It is parents who should really bear the blame for television's damage, Winn states, because it is parents who permit the broadcasts to penetrate the home. In fact, mothers and fathers encourage television-viewing, she says, in order to keep children pacified. When Winn writes, "Surely there can be no more insidious a drug than one that you must administer to others in order to achieve an effect for yourself," she has worded her challenge in such a way as to make thoughtful parents squirm.

The fears of Marie Winn about the disservices television may be rendering the young are amplified by other Media Snobs. One wide-spread apprehension is that television content is coming to replace the real world in the perceptions of children. Frank Mankiewicz, writing in *Remote Control,* says that youngsters "are enticed into believing that what they see on television is what they'll get in real life." The replacement of reality by television fantasy was supposedly what unbalanced young Ronnie Zamora. "Human beings whose primal impressions come from a machine—it's the first time in history that this has occurred," laments Rose Goldsen. Dr. Benjamin Spock, the renowned child care expert, may have personally confronted this problem when he brought his grandchildren to New York City for a tour of the Bronx Zoo and the Museum of Modern Art. Television's images were more fetching to the children than the sights, and he was unable to dislodge them from the hotel room: "I couldn't get them away from the damn TV set. It made me sick."

The video fantasy world which so thoroughly captivates the young is riddled with violence—not only in the Saturday morning cartoons but also in the weekday action/adventure shows which children select for themselves. Although some Media Snobs claim that this phantasia of brawlings and shootings works to heighten tolerance of real-world violence, and "desensitizes" children to atrocity, other critics see it having the opposite effect, stimulating the child to aggress. Either way, it was the issue of violence which galvanized the Parent Teachers Association to mount a national campaign in the latter part of the '70s against television bloodshed. By publicly pointing their finger at violent programs like *Kojak,* the PTA elevated many Americans' awareness of video mayhem, and renewed concern about the lessons learned from it.

Yet another issue, one that has turned many Media Snobs into firebrands, is that of advertising aimed at children. Pliant youngsters, lacking the ability to distinguish between a fantasy program and an advertising pitch, need to be protected from the badgering commercials; this is what Peggy Charren, president of the most influential Snobbish pressure group, Action for Children's Television, declares. Deceptive advertisements for toys are bad enough, she argues, but the sugared cereal and candy pushed by children's commercials do actual physical damage. "I don't expect television ever to be perfect," Charren allows, "but at least it shouldn't hurt children."

Not only is television said to be promoting aggression and cultivating avarice, but other Snobs believe it is crippling children's development by downplaying the ability to read and think clearly. Taking literacy and rationality to be the hallmarks of civilization, Snobs see these attributes threatened by the visual medium which occupies so much of children's time. According to Rose Goldsen, even the highly regarded program *Sesame Street* does harm to the cause of literacy and the love of books. "The 'Sesame Street' curriculum is teaching the culture of the midway impressed into the service of selling products and ideas. It is a curriculum that has nothing to do with books or with the culture of books and reading," her salvo goes. "The daily dramas on 'Sesame Street' have never featured anyone absorbed in a book, laughing or crying over a book, or so gripped by a book that he cannot bring himself to set it aside."

The various forebodings of Media Snobs regarding what *Variety* calls "kid vid" do receive some support from the few objective studies which have been done on the viewing behavior of the young. The highest regarded and still the most frequently cited examination of children's television was published in 1961 by the communications scholar Wilbur Schramm under the title *Television in the Lives of Our Children.* Schramm and his colleagues at Stanford University executed a total of eleven studies over two years involving 6,000 youngsters and 2,000 parents; they were also able to find a pair of towns to study in Canada which were comparable in all ways except that one had television and the other did not. The other major research effort into children's viewing appeared in print in 1972 as part of the Report to the Surgeon General on Television and Social Behavior. Its principal author, Jack Lyle, had been one of Schramm's coauthors ten years before, and in several respects the second study, which tapped 1,500 first-, sixth-, and tenth-graders, was an update of the first.

Both the 1960 study and the 1970 one documented what is obvious to every parent: children like television, and because they like it, they spend a lot of time with it. First-graders and sixth-graders passed about two-and-a-half hours each day on the average with their sets in 1960; by 1970 this span had lengthened by an hour for both groups, to three-and-a-half. Viewing time has probably plateaued at the second figure, although it may still be climbing for preschoolers spending the day at home. Adding in weekend viewing, the typical sum is now roughly twenty-three hours a week for first-graders, and thirty hours a week for sixth-graders.

After the sixth grade, however, viewing time begins to fall off, according to Schramm and Lyle. Teenagers, involved in school activities and the other sex, watch less than anyone else; viewing does not increase again until people have passed on into their twenties. Thus it's in early childhood when youngsters are most inundated by television.

What children prefer to see during the hours they are viewing varies according to age and sex, although as they revealed to Schramm and a decade later to Lyle there is sure to be a high proportion of violence in the diet. Preschool children want above all the blows and boffos of cartoons; listed next are situation comedies, with non-cartoon children's programs a distant third. Nine out of ten three-year-olds can identify Fred Flintstone, Lyle learned. First girls and then boys come to favor situation comedies more, and by the second grade this type of program is the most appreciated. The growing child turns more and more to adult shows, and in sixth grade 80 percent of the viewing is of programs intended for grown-ups.

Those who argue for "high standards" in programming will be disappointed to learn that the major change between 1960 and 1970 in children's viewing was the eradication of class distinctions. Lyle reported that status differences in amount of viewing time and in program choices, which were conspicuous when Schramm did his research, had dwindled over the decade. As the years passed there had been greater and greater convergence of viewing behavior; children from families high on the socioeconomic ladder had grown to have the same preferences as those at the low end. By 1978 George Comstock could observe, "The most striking feature of children's viewing is its lack of variation by household characteristics." This development parallels a similar trend toward convergence of viewing practices among the different strata of American adults.

Sometimes children's viewing further resembles their parents' in that they can be doing other things simultaneously. About 80 percent of first-graders say that their attention may well wander to different activities while the set is on. But many times children are staring deeply into the picture tube, entranced to a degree that can only make a mother or father uneasy. A boy from Lake Forest, Illinois, reports, "Sometimes when I watch an exciting show, I don't blink my eyes once. When I close them after the show, they hurt hard."

Adding to Young Minds

The fears of Media Snobs are premised on the notion that television primarily shoots improprieties into young, unformed minds. "Television has profoundly affected the ways in which members of the human race learn to become human beings," ominously states George Gerbner, the creator of the Violence Profile. Concentration solely on the medium's teachings to the exclusion of other aspects of television which are much more central to youngsters' actual viewing experiences can be the special obsession of Snobs.

To some degree television does put information into young brains. Every mother and father hears phrases and witnesses behavior whose origins could be nowhere else but the cathode-ray tube. In Lyle's 1970 survey, nine out of ten mothers felt that their first-graders had been learning things from television. And children themselves agree, telling poll-takers that television is a regular source of information for them. "The least contestable generalization about the effects of television on young persons," George Comstock says in summarizing three decades of media studies, "is that they learn from the medium."

Children are bound to absorb somewhat more from television per hour of contact than adults do; they are straining to learn all they can about the world they are entering, while adults already have a passing familiarity with it. But how much learning actually goes on, and of what, and by which children are questions that remain to be answered. If we are going to see the whole picture, we'll also have to know how learning from television compares to learning from other sources. And how children's use of television for instruction compares to their use of it for other purposes.

Of all children, it's preschoolers who are most likely to pick up some things from their time with television. They are, after all, the ones who have the most to learn, being the newest arrivals. Their genetic inheritance may have established a fraction of their behavior and outlooks, but everything else must be learned fresh. Yet for all their need, the environment from which they can find clues about the real world is largely limited to the home—and to the television set. Hour after hour they sit and view, for a total time that far exceeds that of their older brothers and sisters.

One thing they learn is some vocabulary. The toddler who says, "frosted flakes," "rub out," "Triple Crown," and "Yabba dabba doo" has been an apt pupil. Television brings much more varied language into the home than any one set of parents normally uses, and the child at this stage is an avaricious language-learner, developing the prime tool of subsequent social life.

Preschoolers also learn items from television in the area of what researchers call "general knowledge"—basic operating information about what the world consists of and how it works. Can boats fly in the sky, for example, and what are keys for? Does the United States have a president or a king? Most preschoolers know.

Much of what it takes an individual to function well in society is not knowledge which can be put into words, but a sense of the nonverbal patterns that guide conduct—who shakes hands with whom, and when; how pets are to be treated; how amazement is expressed; how close people stand to each other when talking; and on and on. This information too is gathered by preschoolers from television. But, as in other areas of knowledge, learning is far more likely to occur when real-world information is lacking. For instance, in the television era rural five-year-olds know how to behave in taxis long before they have seen a real one, much less taken a metered ride. But the same rural five-year-old will still laugh at a television actor's poorly played pretense of milking a cow.

If any television show is sure to teach preschoolers, it's *Sesame Street,* which has been produced by the Children's Television Workshop and offered through public broadcasting since 1969. The program is the end product of highly motivated production personnel (many of whom were hired away from CBS's *Captain Kangaroo*), skillful on-camera performers (ranging from Rita Moreno to Miss Piggy, from Big Bird to Bill Cosby), and an open-minded and capable research staff. It is not this way with other television shows, but learning from *Sesame Street* is actively encouraged by parents, who appreciate the lessons on language and numbers, as well as on how to behave considerately. A senior researcher at CTW explains why the show potentially can instruct the preschool set: "The need to direct one's attention and filter out irrelevant stimuli, the need to hold temporally separate examples in one's mind for purposes of comparison, the need to distinguish essential attributes from irrelevant ones are all requirements of concept-learning which place enormous demands on the small child's capabilities. But television can do all this for him. The television world is a limited world, where nothing happens at random. Because this environment is so controlled, many of the ordinary obstacles to learning are removed." A television presentation, she goes on to say, can shift back and forth from reality to fantasy, and thus is compatible with the young child's thinking processes.

She takes pains to point out that there are two active parties involved in this process—the communicators at CTW and the young viewers. The children are not simply passive recipients of what the *Sesame Street* producers decide to transmit, but are active in the system, accepting or rejecting material on the basis of their desires and abilities. She says, "The child's manner of processing television material is a function of his level of cognitive development and his intellectual or emotional needs at the moment, and the child does not take in whatever the TV has to dish out. The relationship is not one way; it is interactive."

But in spite of the recognition by *Sesame Street* personnel of the essentially two-way nature of television, and in spite of their considerable research and production talents as well as the praiseworthy nature of their mission, and in spite of the almost universal acclamation of the series by adults, the unhappy truth of the matter is that *Sesame Street* does not teach children much. The original examination of the program's teaching, carried out by the Educational Testing Service of Princeton, New Jersey, at the close of the first two seasons, in 1970 and 1971, pronounced it a success; children who viewed had learned language and number skills which would stand them in good stead, the ETS reported. Public opinion still holds this conclusion to be gospel. But a reanalysis of the testing data was undertaken in 1975 by the Russell Sage Foundation of New York as part of a larger project to evaluate social science methodology, and the findings were disheartening. Released in book form by Thomas Cook and his coauthors as *"Sesame Street"* *Revisited,* the 1975 reappraisal said, "We believe that the evidence we have examined casts reasonable doubt about whether *Sesame Street* was causing as large and as generalized learning gains in 1970 and in 1971 as were attributed to the program on the basis of past evaluation."

What the authors of *"Sesame Street" Revisited* pounced on was the fact that the children who had recorded the greatest gains in the earlier tests were those who had received special encouragement to view. This group had the benefit of home visits by educational specialists and colorful supplementary materials, plus the proud attention of their parents. Cook came to feel that the higher test scores were not the result of the CTW broadcasts so much as of the exceptional show of adult interest and support. Almost any four-year-old, when steadfastly concentrated on by grown-ups, will try to be obliging, no matter what the purpose.

As Cook and his research team began to talk and write about their surprise conclusion—that of minute effects at best for typical, unencouraged viewers of *Sesame Street*—they found their message was met with great resistance. The press carried little mention of the 1975 book, and when Cook talked to parent groups he often faced open hostility. In a period of social turbulence, *Sesame Street* had become a rock for uneasy mothers and fathers. The show had laudable intentions, was morally correct, and if the kids were going to watch television, why not this in lieu of the cartoons? *Sesame Street* had become such a darling of parents that they assumed small educational gains on the part of their three-year-old viewers spoke of large gains, or that any and all gains were due to that program alone. Cook did not have glad tidings for them: preschoolers were exercising their rights as viewers to slough off many of the show's lessons. The amount learned was small enough to raise the question of whether *Sesame Street* was worth the expense and effort that had gone into it.

The very best that television has to offer preschoolers accomplishes little by way of instruction, and the rest of the programming does even less. Compared to the volume of hours the toddlers spend gazing into their sets, the amount of learning is very small.

And the rate of learning decreases as the child ages. It's only reasonable that this should be so: as young brains slowly fill up with real-world knowledge, they grow less needy as well as more discriminating about types of information. Books and schools come to be ampler sources when the hours of contact with television decline. Most school-age children recognize a difference between fantasies and reality material, and they know that television excels in the former and that schools, by and large, deal in the latter. Aimee Dorr Leifer of Harvard University has demonstrated that by the age of six the young video-sophisticate knows "the stories are pretend." Only five percent of Lyle's first-graders thought that the people on television were exactly like the people they knew in real life. As the years pass the youngster usually finds television to be less and less informative; the percentage who think they are learning from television drops twenty points between first and eighth grades.

Parents overemphasize the instructive aspects of television because of their personal reservations about what appears to be a competing and untested agent of socialization in their children's lives. No parent wants to abdicate to a machine, much less to commercially motivated programming. Mothers and fathers don't have to look hard to see that some of that content is being picked up; it is a large enough amount to lead a parent to think that children are being influenced more than they actually are. But the vast bulk of what children learn comes from other sources which provide better and more appropriate reality material—first home life and playmates, then also reading and school. Columbia University professor Herbert Gans tells the real story: "Television and other media do not play that large a role in most children's lives: the actions and attitudes they learn from parents and peers are far more important." The way Mommy acts with Daddy, the deportment of siblings and relatives, the interactions of playmates—these are the truly valuable and verifiable lessons in how the world works.

Whenever television does manage to teach, it usually happens this way, according to George Comstock, who has sorted through all the literature on children's learning from television: "The media provide information which young persons submit to others, including parents, for comment and evaluation; and out of this process, in which interpersonal influence is strong, attitudes and opinion are formed." When a child recites something from television to a parent, the mother or father may feel the child is exhibiting a bit of learning, good or bad, but the child knows that learning is not over until the parent has responded. It's the real world which remains children's touchstone, despite the fears of Media Snobs.

The fact that television isn't much good at teaching children was first substantiated through Wilbur Schramm's multi-pronged investigation in 1960. When he contrasted knowledge test scores from youngsters in Teletown (the Canadian town which could receive television signals) to those in Radiotown (the other Canadian town, devoid of television, which had been selected for purposes of comparison), he discovered that the television-exposed first-graders knew more, but that very quickly the differences evened out and the students without television caught up. The older children were equally knowledgeable whether television had been a presence in their lives or not. What youngsters need to know they might learn sooner from television, but they could just as well learn it a bit later elsewhere. Instruction from television was not of great significance, Schramm's study related.

Other studies done at that time by Schramm and his research team from Stanford University revealed that, on the rare times when learning does occur, it is casual and inadvertent on the child's part. The young person does not think of television as a teacher or himself as a student when viewing, and thus that is not the mode of learning that goes on; what commonly happens is that something of interest comes up in the course of a show and the child will latch on to that one item. "Most of a child's learning from television," said Schramm, "is incidental learning." Schramm would have no difficulty appreciating a personal story from Isaac Asimov, the world-famous science fiction writer. Asimov had received a large silver object as a gift, and his nine-year-old son had promptly identified it as a champagne bucket. "How do you know about champagne buckets?" the writer asked. His worldly child replied, "Oh, I see them all the time on *The Three Stooges*."

Not turning to television for instruction, children will carefully avoid any program that seems too didactic. In one survey Schramm learned from his young respondents that they virtually never watched educational, public affairs, or news programs, even if the shows were especially tailored for them. Instead, Schramm stated with finality, "Practically all of a child's use of television is a quest for entertainment."

From a child's point of view, that is the beginning and end of the matter. Entertainment, not information—fantasy, not reality—is what he demands from the medium. Those who arrange for children's television realize this, and if they are honest, confess it. Mary Alice Dwyer, NBC Entertainment's vice-president of children's programming, says, "The nature of network television is that we are an entertainment medium, not an educational medium." Even the executive producer of

Sesame Street, David Connell, had to recognize these priorities if he was to keep the young audience from slipping away and the learning scores from being even lower: "Our purpose is both to instruct and to entertain, but always to entertain."

Before we turn to the content that children really appreciate on television, and see what it does for them, there's another instructional aspect of the medium to be scrutinized—those troublesome commercials.

Children's Commercials

It's true for adults, and it's true for children too: exposure to commercials is the price paid for the programming seen. There is no question that some of those advertising messages do get inside children's heads. A full 90 percent of mothers report that their preschoolers sometimes ask for advertised products. And a majority of these mothers say their offspring will occasionally break out in a commercial jingle. When a third-grade video-generation class in Connecticut was asked to spell "relief," more than half of them wrote "R-O-L-A-I-D-S."

This is what commercials are designed to do—to penetrate and influence young brains. Just as they are for parents, the messages for children will be pitched to deep-lying needs and longings. Toys and sugared foods can be sold by appealing to the youngster's desire to be the first, the best, the most popular, and so forth. Advertisers learn how to engage children's attention through proprietary research done by special companies which test and interview children. On behalf of their clients, firms like the Child Research Service of Rutherford, New Jersey, will poke into children's minds through psychological projective techniques such as role-playing, drawing, and whispering. Marketing strategies result: Texaco determined it could sell more gas if nagging children directed their parents into filling stations where toy fire trucks were given out with each fill-up.

In a society where children are hallowed and commerce is not, the hucstering of goods by seductive television advertisements is not likely to be appreciated. When deceptively advertised toys are singled out, or parents learn (as they did from a University of Georgia study) that if a cardboard cereal box is eaten with raisins and milk it will be more nutritious than the cereal within, then resentments are bound to rise. There is more than a taint of the sinister and underhanded to the business of selling to children through television.

Reaction to children's commercials has come from several reform-minded groups, most prominent among them Peggy Charren's Action for Children's Television. The essential question, as put by a spokesman for ACT, is this: "Should special protection be provided to insulate children from direct advertising designed to stimulate their consumptive desires so that they would become active lobbyists for the merchandiser within the family?" The kind of "special protection" ACT proposes is an absolute prohibition on all commercials aimed at children. Powerful federal agencies have proven receptive to this radical proposal.

Over the 1970s ACT marshalled considerable support for its point of view. In 1973 the FCC received 55,000 letters commenting on television, only thirty-four of which complained about children's commercials; but in 1974 ACT precipitated an outpouring of 100,000 letters strongly decrying advertising practices on children's television. In the late '70s the scene shifted to the FTC, which held a series of public hearings to explore whether or not the commercials should be banned. The FTC staff believed so, and in a 1978 report urged a total ban on all advertisements directed at children under the age of eight, because they are "too young to understand the selling purpose of, or otherwise comprehend or evaluate, the advertising." FTC chairman Michael Pertschuk revealed himself to be sympathetic, referring to the "moral myopia" of children's advertisers. For this show of partiality the advertising industry managed to have Pertschuk legally removed from taking part in the FTC deliberations over the matter. The advertisers' point of view was that if a product could legally be sold to children, it could legally be advertised. They also insisted that the FTC was trying to butt into family concerns, since it was the parents' role to control their youngsters' purchases.

Broadcasters and advertisers additionally stressed that the abolition of children's advertising would mean the end of children's television, or close to it. The half-billion dollars that advertisers annually put into children's commercials sustains the shows for young people; if that money were choked off, the programs would die also. The economic realities are no different than for adult television.

The battle swayed back and forth, but in the end it was the reformers who were beaten back and the advertisers who were the victors. The election of the Reagan Administration signalled a change in the regulatory climate, and on October 1, 1981, the FTC officially ended its consideration of the matter.

Children might not need as much protection as the FTC was contemplating, for the gullibility of small viewers is far less than Media Snobs assert. Social scientists Mariann and Charles Winick had hundreds of children observed in the act of viewing, and they report in their book *The Television Experience: What Children See,* "One generalization that emerged very clearly was the very considerable awareness of commercials as something apart from programs, at all age levels, even the very young. The youngest child seen in this study, two years old, left the room regularly every time a commercial was shown." Parents perceive their offspring becoming discriminating at an early age: 65 percent say their three-year-olds know the difference between programs and commercials. For all children aged three to ten this distinction is recognized by 89 percent. As children age they rapidly become skeptical about the advertising messages; the four-year-old who greets a commercial positively is succeeded by the nine-year-old who is openly disdainful. Growing resistance is revealed in the degree of attention paid to commercials, which drops from 50 percent of the time for five-to-seven-year-olds to 33 percent for eleven-to-twelve-year-olds. The waves of commercials, season after season and year after year, do not shape capitalist stooges but rather highly sophisticated consumers.

Of the tens of thousands of commercials each child sees yearly (the National Science Foundation in its report *Research on the Effects of Television Advertising for Children* calculated the number to be 20,000, but this seems on the low side), only the smallest number move the child to turn around and importune his parents. There is some evidence that in homes where the level of hostility is already high, the child will be more likely to besiege the adults in his life for advertised products; those children learn that such pleas can be another volley in the ongoing battle. But the average child has a very limited number of requests, and parents do what they can to oblige him, buying two-thirds of the cereals asked for, and one-third of the toys.

To see the relationship of advertiser and child in full perspective, it should be noted that many of the purchases bring the youngster a great deal of pleasure, just as many adult purchases do. The candy is as tasty as promised, or the toy becomes an often-used enhancer of the child's activities. For child as well as for parent, the relation between producer and consumer is usually a reciprocal one, and beneficial to both parties. Parents' antipathy toward children's advertising may stem in part from their sense that their boys and girls are growing up just like them, making indulgent but yet gratifying choices of products. Jonathan Price says in his *The Best Thing on TV,* "Why do parents, who buy tires, wines, film, cards, transmissions, cars they first saw on TV, get upset when children ask for something *they* saw on TV? Maybe it's like a bad habit. Kids doing it reminds us that *we* do it."

However, parental attitudes toward advertising for children are in truth not very negative at all. The majority of parents think it's perfectly all right to have commercials on children's programs—especially if that means the programming will continue. A recent survey of 150 randomly selected parents of children aged two to twelve found that the average level of concern was not high. The nonchalance of mothers and fathers was exhibited in the finding that no statistical correlation existed between dissatisfaction expressed by a parent over the commercials and any attempt to actually monitor their children's viewing. No matter what they may say, parents are not offended enough by the advertisements to do anything. It is possible parents may not be activists on this issue because they sense there are some benefits from this sort of consumer socialization, or at least few losses.

Cartoons

The instructional aspects of television for children, which are the focal points of so much adult discussion, count for little with young people. What television means to them is fantasy and more fantasy. "When children talk about the gratifications they get from television, the fantasy gratifications come out first and in much greater number," Wilbur Schramm related. "When they list favorite programs, fantasy types of programs are likely to outnumber reality programs by a ratio of twenty to one."

The networks have always been ready to oblige this demand. "We're storytellers, not teachers," said NBC's vice-president for children's programs, George Heinemann, at a National Association of Broadcasters conference on children's television. "Leave the teaching to the teachers in the classrooms," he remarked, voicing the feelings of many programmers present. In search of the largest possible audiences, broadcasters know that it is foolhardy to deliver to children shows they don't really care for. "As programmers, we're not successful unless most of the kids are watching us," said Heinemann's counterpart at ABC, Squire D. Rushnell, about Saturday morning shows. Broadcasting is not a complex business at bottom.

Children's call for fantasy should not be thought of as something trifling or dismissable. It's not for them, and not for the networks, and it shouldn't be for us either. Highly important psychological needs of youngsters are met through television fantasy. Childhood is a period when the energy and impulsiveness of the new personality has to come into conformity with the dictates of society so that, later on, the new entrant will be able to play a productive adult role, and human community will endure. "The child has to learn to live in a family, ruled by parents who sometimes seem to him to be inconsistent or unfair. He has to learn to play the role expected of a child, and to bend his behavior to the patterns taught him as part of his socialization," explained Schramm.

The child's behavior is not bent without some resistance on his part. Even in the most considerate of families, the best efforts of parents are greeted with early willfulness on the part of two-year-olds and continuing opposition on through the rebelliousness of adolescent years. It's probably just as well that it is this way, for a flaccid personality is to no one's liking; nevertheless there arc unavoidable resentments that build up on both sides. The parent has no doubt adopted personal ways to handle such anger efficiently and harmlessly, but the child has to learn from scratch the socially acceptable behavior which will serve to discharge animosity. Seymour Feshbach, the psychology professor who did the field study on adolescent aggression and television fantasy, says that one of the greatest developmental lessons of childhood is learning how to control retaliatory impulses, and how to direct them into proper channels.

If resentment for the impositions of parents is one source of youthful hostility, another is nothing more nor less than the simple immaturity of the child. Children can't accomplish what they would like to and often become frustrated as a result. Psychologist Jerome Lopiparo, writing in the journal *Intellect,* ventures, "Probably the most important reason aggression must be a part of a child's life is that it helps him cope with his feelings of powerlessness. Children do not feel powerful. They do not feel that they can affect change either in their own lives or in the lives of others. If they want to do something, they must ask an adult; if they want to have something changed, they have to hope adults feel the same way, for that is the only way it is going to happen."

That is the problem; Lopiparo next goes on to say what most children have found to be an acceptable solution to their aggressive feelings: "It is for this reason that children are drawn to TV violence. Many of the frustrations they feel can be very effectively worked out via the TV screen. It is safe, the person you are attacking cannot retaliate, you can be a hero or a villain with just a flip of the dial, and, most important, you can experience that elusive feeling of *power*."

"What kind of TV program offers children the most fascination? Which ones do they sit looking at in excited wonder?" Lopiparo asks rhetorically, and he answers, "It seems that the more gore the program has, the more they like it." Gory programs usher children into an otherwise forbidden world where they can vicariously vent their frustrations and hostilities. Seething resentments and impulses toward unspeakable cruelties can be spent in a way which is not absolutely chastised by society, and which is—more importantly, from the small child's point of view—totally free of any chance for hurtful retaliation. What Feshbach discovered about the way television fantasies serve adolescents also applies to younger children.

The children's shows with the most gore, and thus the shows that are the most sought out, are violent cartoons, televised predominantly on Saturday mornings. In 1973 Dean Burch, then Chairman of the FCC, asked, "Is Mickey Mouse running over an opponent with a steam engine and the opponent afterward puffing up to normality and running off—is that 'violence'? I don't know." He may have feigned ignorance, but we can answer for him—it certainly is violence. Hitting, tossing, choking, bombing, smacking, flattening, igniting, exploding—this is regulation violent action on cartoons, and it is a true obfuscation of the matter to express doubts that it exists. To hold their audience cartoons feature a violent episode at least once every two minutes, by careful count.

It is true that the logical consequences of violence are not depicted on the cartoons. One rapidly maturing eight-year-old reported, "In cartoons, they never have the character die. He just gets all black and blown-up—and then they make like it doesn't hurt." If the cartoons did show a realistic aftermath, it would destroy most children's pleasure in them. What children want from the shows is the chance to aggress vicariously without being burdened by any repercussions. The real world is where aggression produces the responses—either blows or self-blows in the form of guilt—which usually disallow its expression; it's the real world that children want momentarily to get away from when viewing cartoons.

Although cartoons are the favorite programs of the very young, Saturday morning does not belong exclusively to them, nor is that even when they do most of their viewing. Fifty-one percent of the viewers before noon on Saturdays are above the age of twelve. And Saturday mornings make up only 16 percent of the total weekly viewing time of children aged two to eleven; most of their viewing is done in the late afternoon and early evening each weekday. But it does no disservice to the programming and advertising realities to say that the Saturday morning cartoons are the essence of television for the young.

Why do the Saturday morning shows consist of animation rather than actors? Cartoons have special attributes which make them ideal for their audience and purpose. Young minds are not yet skilled at pulling out from a scene the details that are important, but cartoons can reduce the amount communicated to a few essentials. Also, the drawings can convey much that is consonant with the prelogical thinking of children but would be difficult to depict using real actors and settings. Characters can fly, rip up mountains, part cities, turn houses into automobiles and automobiles into vaulting poles, and do all kinds of wondrous things which would be next to impossible to accomplish by any other production method. Animals can speak, motorcycles can have personalities, buildings can swell and shrink, movement can be slowed down or speeded up, and on and on. Cartoons establish a fantasy world which is not too overlapping with the real world, but of sufficient distance where normally untoward things can happen without the fantasizing child being led to feel uneasy or guilty.

Among the reasons that cartoons are the standard mode of presentation on Saturday mornings is that they have become economical to produce. When studios like Walt Disney made cartoons for movie theaters in the '30s and '40s, the process was comparatively expensive, calling for twelve drawings per second of film. But nowadays, in what is known as "limited animation," as few as two drawings are used, and costs are proportionately lower. People inside and outside the animation business are scornful of these changes, seeing them as a cheapening of the art, but young viewers do not seem to mind. Elaborate cartoons were appropriate for audiences with a large proportion of adults, but the very young prize simplicity. It does not bother them at all if backgrounds are fixed and characters are positioned except for flapping jaws or cycling limbs.

By forcing production costs down, the networks have been able to make Saturday morning cartoons very profitable for themselves. Of the half billion dollars the advertisers of candy and toys pay out, much of it stays in network pockets. One way the profitability of Saturday mornings is increased is through the use of reruns—a careful viewer of Saturday mornings can plan with the use of a program guide a short course in the history of cartoons, since few that come into the possession of the networks ever go into permanent retirement. Repetition does not bother the young audience; in fact, knowing what will happen seems to increase their enjoyment. In the research project done by the Winicks, one child watching television with a note-taking observer exclaimed gleefully at the start of a cartoon, "I've seen this one a hundred times," and he may well have.

Television fantasies do accomplish for children what children want them to—they help ease out tension and hostility from young minds. When Lyle interviewed children in 1970 about why they chose to watch television, they told him that relaxation was a prime reason. In order to relax the youngsters were more likely to turn to their television sets than to music, movies, reading, games, sports, conversation, or solitude. Maren Winn disparagingly quotes a mother's report which most children and parents would accept more tolerantly: "When Davy gets home from school the TV helps him relax. He's able to turn himself off a little bit with it, in a way."

That the cartoons are particularly good at stripping young minds of tensions and resentments is suggested in the findings of one experiment where children aged four to twelve were shown both violent Vietnam War footage and violent cartoons. One hundred percent of the young viewers recognized the battlefield film as violent, but the great majority (73 percent of those four to eight and 83 percent of nine-to-twelve-year-olds) thought that the cartoon fantasies were *not* violent. This is reminiscent of similar findings among adult viewers, who testify that their favorite action/adventure shows are nonviolent. Children do not find animated aggression to be violent because the fantasies have come and gone through their minds, picking up some of their pent-up animosity and frustration on the way. The cartoon fantasies do not linger and fester, galling the young viewer, as the real-life war scenes did.

Evidence that the cartoons have the opposite effect, and stir up rather than relax, is weak at best. Roughly a dozen experimental studies exist on the specific topic of television cartoons and children's aggression, and a recent overview of them concludes that, no matter what the experimenter's biases might have been, the studies cannot be used to indict the cartoons. In this review of the literature, Professor Walter Hapkiewicz said that some experimenters had elicited aggressive display toward inanimate objccts, but none had found any toward other humans. Representative were two experiments Hapkiewicz had done himself, in which he had shown violent Mickey Mouse and Woody Woodpecker cartoons to children six to ten years old; in neither experiment was subsequent play among the cartoon-viewing children any more boisterous than the play among other children who had seen an innocuous cartoon or no cartoon at all.

Media Snobs find it difficult to believe that the cartoon violence does not have negative effects. Senator John Pastore insisted, "Violence on television and the effect that it has upon young minds is not a figment of my imagination. All you have to do is watch the tube and you can understand it." It seems so simple—an adult looks at the violence being broadcast and assumes it's being inserted into viewing youngsters' minds. But Pastore and others are deceiving themselves, for the "instruction" is indeed a figment and misperception. Pastore's mistake is the mistake of most adult observers of viewing children: they study the young people with notions of instruction and rectitude foremost in their minds, forgetting for the moment that some of what children must do to maintain their psychological equilibriums is to rid themselves of harbored tension and resentments. The main function of television for children is the same as for adults: it does not put things into brains so much as take things out.

The misinterpretation of children's television fantasies by Media Snobs, and the subsequent misguidance of parents and other interested adults, bothered psychologist Jerome Lopiparo: "Most critics have failed to advise the public of the positive aspects of media aggression for the vast majority of children and adults who are not on the brink of violent behavior, but who, nonetheless, need the vicarious release these programs provide. Perhaps the bulk of the difficulty lies in our unwillingness to accept the premise that within all of us—child and adult alike—there reposes a potential for violence, TV or no TV. Once this is accepted, we may be more receptive to the notion that the *expression* of this aggression, whether via fantasy or outright overt behavior, is not only normal but, in many respects, quite beneficial. What we are then left with is the possibility that the aggressiveness our children watch on TV may actually be *reducing* overt expressions of violence, rather than increasing them, as the critics would have us believe."

On Balance

If the main short-term effect of television upon children is the removal of some of the mental pressures that unavoidably come with being raised up into human society, questions about long-term effects still remain. Media Snobs wonder if a dangerous, epochal turn of events is under way whose innocent agents are the impressionable minds of the young—to wit, the replacement of a time-honored mode of communication by an upstart, crass one. Snobs fear that reading is succumbing to viewing.

Marie Winn suggests, "Indeed, the mental differences demanded by the television experience may cause children who have logged thousands of hours in front of the set to enter the reading world more superficially, more impatiently, more vaguely." The best evidence in support of the view that reading abilities are on the decline comes from the standardized College Board tests given at the end of a youth's spell with the schoolroom and television curricula. The average scores have been dropping steadily since 1963, falling 50 points in the case of reading and writing skills. Winn remarks, "The decline in scores may clearly be related to the steady increase in television ownership in the United States from 1950 on."

But not all standardized tests offer up the same disquieting picture of eroding skills. In 1965 the state of Iowa administered to all elementary school children the same diagnostic test that had been used with pre-video age children in 1940. The results were notable—not only were the students' abilities at reading and comprehension much higher than in 1940, but so were skills at conceptualizing, handling abstract symbols, and reasoning. In Indiana the statewide reading scores for sixth- and tenth-grade students in 1976 were compared to those of 1944; the findings were of unmistakable improvement for children of the television generation. "The nation as a whole is concerned today's children do not read well," remarked Dr. Roger Farr of Indiana University, where the analysis was conducted. "Any well-done study such as this is evidence to the contrary. Children are reading as well as in 1944–45—actually, reading far better."

Scores on national tests carried out during the 1970s also confirm the unfaltering reading abilities of America's young. Funded by the federal government, National Assessments of Reading were conducted in 1970, 1975, and 1980. They revealed that 13-year-olds and 17-year-olds did not slip at all over the course of the decade, while in the 9-year-old bracket, significant gains were registered. Nine-year-olds at the end of the decade were reading markedly better than 9-year-olds at the beginning.

Why is there such a discrepancy between the College Board results and those of Iowa, Indiana, and the National Assessments of Reading? The problem lies not with the tests themselves; their scores are accurate enough. The difference is mainly accounted for by the nature of the pools of students being examined. The ethnic and income-level composition of the Iowa and Indiana schools, as well as the national school-age population, has remained relatively stable during the testing periods, permitting valid comparisons over time. But College Boards are now taken by many kinds of students who previously did not finish high school, much less aspire to college. In 1975 the College Entrance Examination Board itself convened a special advisory group of 23 experts to discover why the scores had been dropping for over 10 years. Their report, made public in 1977, cited the "notable extension and expansion of educational opportunity in the United States." One member of that advisory group, Ralph Tyler, vice-president of the Center for the Study of Democratic Institutions, explained, "Judging from the data, our efforts to extend educational opportunities are resulting in more young people from minorities and low-income homes finishing high school and entering college, but their educational achievements, although rising, are still lower on the average than those of other students." The sliding College Board scores tell not of a dip in American's reading skills, but of the recent opening of higher education to groups previously discouraged. It's ironic, but the flagging scores reveal social progress, and not the reverse.

The widening pool of test-takers explains much but not all of the drop in SAT scores. A second influence has been unearthed by Landon Jones, author of *Great Expectations: America and the Baby Boom Generation.* Again, it is not television that is the culprit. In Jones's analysis, the students taking the SATs in the 1960s were the firstborn in the post-World War II baby-boom families. Their kid brothers and sisters came along and took the test in the 1970s, and produced lower scores. Why? It's a fact that second- and third-born children tend to score lower on intelligence tests. Birth order, as it influenced the constitution of the outsized baby-boom group of college applicants, was a reason the SAT scores dropped.

Television-viewing is not undermining reading skills among the young. In fact, there is a tendency, first documented by Wilbur Schramm in 1960, for the two activities to go hand-in-hand—the student who is a heavy user of print is prone to being a heavy user of television too. Such students tend to be the most successful and best adjusted, Lyle noted ten years later: "Indeed, children who were high users of both television and books were equally or more likely than their classmates to be among the most active in sports, recreation, and social activities."

The explanation for this is that, in general terms, reading and television-viewing are not competing activities, but complementary ones. When Schramm analyzed his Radiotown and Teletown data he learned that television was taking over the fantasy services previously performed for children by comic books, movies, radio serials, and escape magazines, but it was not cutting into what Schramm called "reality media"—books, newspapers, and non-fantasy magazines. Broadly speaking, reading brings information and adds to young brains, while television brings the visual fantasies that subtract. The well-balanced youngster is one who utilizes both modes of communication.

Whether reality comes to the child indirectly through print, or directly through real-world experiences, television is not displacing it, but is coexisting with it. Although many adults worry that video fantasy is substituting for reality in their children's lives, George Comstock could find no scientific evidence for this in his review of all the pertinent studies: "We must emphasize that a persistent theme throughout this literature is that most children watch television when there is nothing better or more necessary to do. Given a choice between viewing and playing with other children, or between viewing and participating in some organized social activity, the majority of children prefer social interaction. We also find little evidence that television has markedly decreased the amount of time older children spend on homework or other 'necessary' activities."

Although reality is of far greater importance to children than fantasy, for a few hours a day most youngsters will feel the desire to turn to television. At the very start of his seminal book *Television in the Lives of Our Children,* Wilbur Schramm raised the question of whether children sought out video fantasies, or whether broadcasters forced them upon the young audience, with noxious effects as the result. "In a sense the term 'effect' is misleading because it suggests that television 'does something' to children," he wrote on the first page. "Nothing could be further from the fact. It is the children who are the most active in this relationship. It is they who use television rather than the other way around."

Children use television fantasies for therapeutic purposes, as an antidote. Here is how NBC's vice-president for children's programs, George Heinemann, sees it: "You send a child to school five days a week; the teachers beat him over the head and tell him you gotta do this, you gotta do that. My boss says the same thing to me all week. Then he gives me Saturday and Sunday off to relax and recover. Now the child should have the same privilege."

To relax and recover—that is the purpose television serves for children, just as it does for adults. The most striking feature of children's television is not how different it is from adults', but how similar. In both cases the fantasies—which often covertly or overtly deal in aggression— help to reduce the viewers' mental strains by allowing us to indulge in bursts of laughter or vicarious plummeting. Children's minds are very much like ours, and so are their needs. It's only common sense that this should be so, for if their minds were somehow unlike ours in these fundamental respects, it would be hard to believe they could grow up and replace us successfully.

Intuiting the real good that television does for children may be the reason that parents, while ostensibly wary of the medium, are in the last analysis quite accepting of it. Study after study has discovered that parents do very little to influence their offsprings' viewing habits. When asked to say whether their child is better off with television or without it, about three-quarters of American parents will confess "with."

11
Television Heals

The Need for Fantasy

Here's an outline of a story, one of just twelve episodes in the series:

The plot revolves around the retrieval of gold objects from a band of feminist militants. The central character is a man of exceptional attributes—great physical strength, remarkable mental powers, photogenic appearance. A prominent social figure has sent him on the mission to find the band and capture the gold. As members of the audience, we are aware that his instructions have been overheard by a sinister-looking woman.

To discover the stronghold of the militants is no easy task. Tracking them down, the hero and his two companions first stumble upon the hideout of another group of radicals, all males. A bruising fight breaks out, but the hero so decisively bests these men, and so impresses them, that several volunteer to join him in his search after the gold.

The lair of the feminists is finally reached. The hero manages to meet up with the leader of the band outside their stronghold, and although the lovely, spirited woman tries to resist her own impulses, it becomes clear that she is attracted to the hero. Eventually she offers to give him the gold without a struggle. But inside the feminists' camp, the lady who eavesdropped on the hero has appeared out of the blue. She warns the rest of the women that their leader is about to betray them.

A battle ensues between the armed men and the fierce women. Males strike females and females batter males. The fight is prolonged, but in the end the hero and his companions win out. Tragically, the leader of the feminists is killed and dies gazing fondly and sadly into the hero's eyes.

The hero seizes the gold and returns home triumphantly.

Is this an action/adventure show of the sort Hollywood television studios crank out? It is not, at least not originally. The elements of this tale have been around for 2,500 years. It's a story of one of the most popular of ancient Greek heroes, the mortal but godlike Heracles (Hercules to the Romans). Taking the gold from the Amazons was the ninth of the twelve labors of Heracles.

The ancient Greeks had a burning need for this sort of fantasy material. Their culture also provided other kinds of artistic creations to engage their imaginations—sculpture, architecture, painting, and music among them—but nothing was as captivating for the entire population as the myths and legends which over the years evolved into the stage presentations of tragedies and slapstick satyr-plays. The Athenians in particular were devotees of the theater, holding yearly competitions for the most gripping and well-wrought plays. A tragedy such as *Heracles* by the renowned Euripides would play to thousands in the semicircular Theater of Dionysius, and would involve hundreds in the production. Such a community fantasy gave the Athenians an opportunity to act out vicariously their discomforts with an imperfectly understood universe, as well as with their fellow humans and, not the least of it, with themselves. The more the Athenians achieved as they established the conditions under which human talents could flower—and theirs is a record of accomplishment unmatched in history—the more they demanded and prized dramatic fantasies. As their commercial and cultural hegemony spread through the eastern Mediterranean in the 5th century before Christ, their counterbalance seemed to be the fantasies staged in the Theater of Dionysius.

For the Athenians and for us, for all peoples in between, for those who came before and for those who will come after, a yearning for fantasy lies at the core of human existence. Fantasy may well be the best redemption for the arduousness of the real world, the saving grace of human life. In summary, this is how fantasy serves:

Since *homo sapiens* first appeared, the human brain has always allocated most of its energy and ability to negotiation with reality. In fact, the condition of "consciousness"—something which continues to trouble the philosophers and psychologists who ponder it—can be seen as nothing more than the operating mode of the brain when it is dealing with or reflecting upon the real world. During periods of conscious awareness (which generally describes the workday), the brain remains alert and poised. If this wondrous organ was not attentive to what surrounded it, and did not promote prudent behavior, then individuals would have failed in the economic and social realms where success, and thus survival, must be had. But for most humans most of their lives, the mind has permitted enduring instead of subsiding, swimming instead of sinking, and as a result society overall has not gone under.

For all the brain has accomplished in aligning individuals to the real world and keeping each one a functioning member of society, there have always been costs. Just being alert entails a certain degree of tenseness, and the tension does mount up. Most societies, now and in history, offer little opportunity to get rid of this tension during the course of a day, and little chance to discharge the retaliatory feelings that can form in response to the onrush of real-world stimuli. These impulses must be held in check, and so must the urges that well up from more primal regions of the brain. Much of the time human beings have to do what Archie Bunker tried to get Edith to do—they must stifle themselves.

The brain is an exceedingly complex center whose workings even at this late date are still the subject of various interpretations and some controversy. But one thing that's clear is that the brain has always had the capability of holding back impulses and drives which might be damaging to the individual or his reputation, or to the fabric of his society. Inappropriate energy can be repressed for a time, sentenced to a holding tank located as far beneath the level of consciousness as possible. Later, when in repose, the brain will produce a stream of fantasies that spurt from the unconscious and in effect reduce the pressure there.

If individuals are not allowed to dream or day-dream—an effect modern experimenters can accomplish by taking readings of brain waves and bringing subjects back to full consciousness whenever fantasizing begins—then they soon become uncomfortable and, in time, disoriented. The opposite is true too: if individuals are denied real-world stimulation, as when they are placed in a tank of saline water with blindfolds on for a sensory-deprivation experiment, then they also become confused and disoriented after a time. What the human brain seems to demand is a balance of reality and fantasy. Both have always been needed for good mental health.

Fantasizing, and the playing-out of deep-lying mental pressures, has been enhanced throughout history by fantasies which are not formed within the individual's mind but are supplied from outside. Through the ages people have flocked to storytellers for the narrations that had the effect of soothing troubled brains. To judge by the enormity of the audience for Heracles's twelve labors and for *Roots,* for satyr-plays and for *The Beverly Hillbillies,* these supplemental fantasies have had an appeal above and beyond private fantasies. This is because they can more artfully stimulate and more thoroughly discharge repressed emotions. When all goes perfectly, members of the audience will experience the mental purging that Aristotle identified to his fellow Athenians as *catharsis.*

(As an aside, a scholarly debate continues over exactly what Aristotle did write on this count. Media Snobs who want to discredit the catharsis theory of television violence advanced by Seymour Feshbach claim that catharsis was an idea of such little importance to Aristotle that the philosopher barely discussed it in his writings. The little Aristotle did say seemed to limit the concept to the emotions of fear and pity; there is no mention in his *Poetics* of aggressive feelings being vented by drama. But what must be kept in mind is that not all of Aristotle's writings have survived; in particular, the second book of the *Poetics* has been lost. Aristotle announced in his *Politics* that he intended to have a full discussion of *catharsis* in his upcoming writings on poetry, so Aristotelian scholars have concluded that the promised discourse was contained in the lost volume. There Aristotle would have tied the idea to the full range of audience emotions. He would have explained why he had selected the term *catharsis,* whose semantic roots are explicit; it denotes to void, to purge. The voiding of psychological pressures is precisely what dramatic fantasies do.)

The Athenians received their therapeutic dose of fantasy productions —whose tastefulness ranged from the most refined to the most vulgar— in the Theater of Dionysius, and we get the same thing from our television sets. Before our eyes every evening there unreel the fictions and phantoms which are the antidote for daytime strains. To elicit the discharge of reined-in drives, television presents, instead of a rendition of the real world which is much to blame for these repressions, a sort of half-world where violence exists but not pain, sex but not lovelornness, humor but not misery, youth but not age, heroics but not passivity, comradeship but not loneliness, gain but not loss. By being distorted in these ways, televised fantasies can provide after-hours redress for the shortcomings of the real world.

Let's take just sex and violence. Media Snobs serve their interests well by focusing on these, for they are the two essences of television entertainment. In situation comedies, as we've seen, they are cloaked in humor, while in action/adventure shows they are frequently undisguised. The reason these two themes are so prominent in televised fantasies is that they are so prominent in the subconscious repressions of viewers. The real world does not provide all the opportunities people desire for aggressive retribution or sexual contact, but dramatized fantasies offer us a setting to carry out these drives vicariously and without repercussions. Individuals need this harmless outlet and would suffer if it were denied them. Gerhardt Wiebe stated in terms of his explanatory scheme, "Because the very essence of *restorative* messages is their token retaliation against the establishment, the likely effect of well-intentioned attempts by proponents of high standards to 'improve' popular *restorative* content is clear. Let's take out the violence, they say, and substitute a theme of cooperative problem solving. The *restorative* essence is removed and *directive* content is substituted. The psychological utility of the message is altered and its popularity is correspondingly reduced."

By having television entertainment with adequate sex and violence, Americans are nightly able to empty their subconscious; aggressive fantasies produce tranquil minds. Writing a column for the *Wall Street Journal,* Daniel Henninger observed, "Most people use television to relax. They use it to relax because the medium produces restfulness—the mind is made blank, one stares, and in fact the eyes defocus slightly. Indeed the condition of watching television very much resembles the *dyhanas,* or stages, of Buddhist meditation, whose purpose is to achieve a state of equanimity by drawing one's mind away from earthly concerns. One text indicates that during dyhana one might meditate by concentrating on a colored patch. The events that occur between the time we get up and Prime Time are our earthly concerns. Television is our colored patch. Viewing television induces something resembling the pleasant, trance-like qualities of meditation. This is why we watch so much of it."

The restitution that television fantasies provide individuals can be seen if people are moved to tell of the deepest-lying reasons they are aware of (and willing to confess) about why a certain show appeals to them. I teach an adult college course called "The Television Culture," and the first writing assignment is a paper on "Why My Favorite Program Is My Favorite Program." It's a difficult assignment because it's hard to be analytical about one's own behavior, especially when that behavior stems from levels of the brain below the most rational and laudable. Nevertheless, most students in the course take the opportunity to dig into their true relationship with a favored program, and try to articulate the depth and nature of the contact their minds make with the chosen fantasies. Sometimes their accounts illustrate in the extreme the needs that all human beings have. A divorced woman in her early thirties wrote, "I live alone with my two boys. Time and money put dreams at a distance. Participation costs. I long to fortify my belief in family, humanity, and the American dream. Every Thursday night, *The Waltons* reassures." A man in his forties said, "Because I have cable, I can pick up a half dozen or more *MASH* episodes every week. I watch every one I can. It's a world away from my world, one I can enter into and leave my world behind. In the *MASH* world I can laugh, wield a scalpel, and mingle with people I like and need."

Another male in his forties indicated the pleasure he got at the end of the work week from *The Incredible Hulk*: "Usually, on Fridays we as a family go out to eat, but we are always back by seven o'clock so I can watch *The Hulk*. If I see I am going to be late for the start of *The Hulk*, watch out!" Better that David Banner turns into the Hulk than he does; he ended, "*The Incredible Hulk* may not be for everyone but it does me a world of good after a hard week's work. It is an hour well spent for myself." While he broke out of workday restraints and aggressed vicariously as the Hulk, a schoolteacher and housewife found her tension-reliever on Fridays to be another program: "It is Friday afternoon. The final bell has rung and the last of my students has climbed on the bus and headed home for the weekend. I am alone in my classroom, exhausted and ready to collapse at the end of a long week. But I can't collapse yet. There are still lesson plans to write, materials to prepare, kids to pick up at the baby-sitter's, dinner to fix, baths to give, and bedtime stories to read. And after all that, like a light at the end of the tunnel, there is *Dallas*."

Occasionally in "The Television Culture" course there will be a student who says he doesn't watch television. He would be among about six percent of the American population, distributed evenly through all demographic categories, who are not television viewers. A man in his thirties told me that he doubted he saw television two or three times a year; television, he said, reminded him of dying. For a moment I thought he was making the sort of argument about deathliness that Jerry Mander did in *Four Arguments for the Elimination of Television*; I encouraged the student to write out his thinking about his nonviewing. On paper he stated, "I believe this is linked to the fact that I worked in a hospital shortly after high school and my job was to go from room to room, doing vital signs. I think I developed my quirk from this period because for every death I discovered (and, working on an acute care floor, there were many), the television always seemed to be on."

But almost all other Americans experience television as life-maintaining, ridding the mind of toxic stress and animosity. Of all possible ways to compensate for the strains of modern life, televised fantasies are what most of us, most of the time, have found to be the most congenial means. We shouldn't forget that television serves two other purposes too—daytime serials add rather than subtract from minds, as does informational programming—but the greatest part by far of the television experience consists of the evening and weekend fantasies that roll through our minds and clean out the mental debris. We are more receptive by far to this "cleaning-out" function than to any "stuffing-in" attempts. Edmund Carpenter noted, "Although TV showed these things, many believe that Kennedy lived on as a vegetable, and that the moon landings did not transpire. TV is fictive in essence."

Reformers

It's Media Snobs who want to interfere with Americans' election through the Nielsen reports of the fantasy material they need. Usually this meddling amounts to wordy criticism, to little more than calling good medicine bad. Afterward Snobs will feel righteous, the viewers who listen to them will feel slightly chastened, but things will proceed much as they always had. From time to time, however, Snobs have been able to mount a much more serious challenge to television programming, with much more worrisome results.

Following the 1972 Report to the Surgeon General on Television and Social Behavior, the remainder of the decade saw a number of Snobbish sorties. Action for Children's Television became a highly visible proponent of change in programming for youngsters, and Nicholas Johnson's National Citizens Committee for Broadcasting was heard from more and more as time passed. Media Snobbery was most powerfully promoted, however, by two campaigns whose strength derived from the large and wealthy organizations that underwrote them—the American Medical Association (AMA) and the National Congress of Parents and Teachers (PTA). Since the AMA and the PTA claim caretaking responsibility in two vulnerable areas of human life—our health and our children—they were listened to carefully when they proclaimed that some television content was pernicious and should be curtailed. As the nation entered the '80s, demands for thoroughgoing reform were heard from a third sustaining institution—religion, and most particularly the fundamentalist groups sometimes collectively referred to as the Moral Majority.

The president of the AMA, Dr. Richard Palmer, explained why his organization had taken up the cause of violence-free programming: "TV violence is a mental health problem and an environmental issue. If the programming a child is exposed to consists largely of violent content, then his perceptions of the real world may be significantly distorted and his psychological development may be adversely affected." There were two thrusts to the AMA effort. The first was to link up with other anti-media campaigns through financial contributions: in 1977 $100,000 was given to George Gerbner for his work with the Violence Profile, $32,000 to the PTA, and $8,000 to Nicholas Johnson's NCCB. The second thrust was to bring direct pressure to bear against television advertisers. Ten major corporations were asked to review their policies about sponsoring violent shows, and four of them—Sears, Kimberly-Clark, Schlitz Brewing, and General Motors—agreed to shift their commercials.

The PTA went about things a little differently. Although a 1970 pamphlet from the organization titled *Mass Media and the PTA* had stated, "The PTA is not a censor of books, magazines, films, plays, radio, or television. It respects the right of every person and every group to judge publications and productions and choose their own intellectual and entertainment fare," by 1975 the organization's perspective had shifted. That year a resolution was passed at the PTA convention demanding that broadcasters reduce the amount of violence on both child and adult programs. To put force behind this, a comprehensive effort was begun by the organization in 1976, one which included marathon public hearings in eight cities. The people who stepped forward to testify generally agreed that violence on television inspires imitation; that television presents a distorted picture of the world; and that television undermines the quality of life. Armed with fifty thousand pages of mostly vitriolic anti-television testimony, the PTA presented its case to the networks, the federal government, and television advertisers.

What PTA officials called their first "Action Plan" was in effect for a six-month period from July 1, 1977, to January 2, 1978, during which time the networks were said to be "on probation." The 6.5 million members of the PTA were asked to flood broadcasters with letters about objectionable programming. Across the nation members monitored local stations and reported back to the National PTA's "TV Action Center." Evaluations of programs were compiled in the first *TV Program Review Guide,* which rated the network shows for wholesomeness: the best show on television, according to the PTA, was *Donnie and Marie,* while the most violent, as we know, was *Kojak.* Since then revised *TV Program Review Guides* have continued to appear, with the scope being widened a bit each time, the PTA says, "so that we are keeping a critical eye on *all kinds* of offensive programming (violence included), and striving to achieve a better, overall quality of TV shows."

The PTA also adopted the AMA's tactic of going for the jugular in the television business—the flow of advertising dollars. In the spring of 1978 the organization called fifteen major advertisers to a meeting in Chicago, where pressure was put on them to refrain from sponsoring shows that dealt in violence or sex. Afterward Sears announced that it would pull its commercials from *Three's Company, Charlie's Angels, Starsky and Hutch, Baretta,* and *Hawaii Five-O.* Maneuvers like this sent bone-deep chills through the broadcasting industry. The extent of the fears can be gauged by the response of Roy Danish, head of the Television Information Office of the National Association of Broadcasters. During the PTA's series of hearings Danish managed to have himself put on the docket in Kansas City, where he pleaded, "I predict that if you make advertisers your go-betweens, you will drain the vitality out of television and you will also invite far more mischief by others whose political or social views you may find distasteful. Broadcasting had its brush with this kind of boycott in the Fifties and the results were shameful and long-lasting. Please, let's not open that Pandora's box again."

But opened again it was when on February 2, 1981, the Rev. Jerry Falwell, president of Moral Majority Inc., joined with the head of the National Federation for Decency, the Rev. Donald Wildmon, in forming the Coalition for Better Television (CBTV). Under Wildmon's leadership the Coalition aimed to identify the more morally wayward shows on television and to press for their elimination. One television personality Wildmon disapproved of was Phil Donahue, whom he termed a "sex activist," and whose daily show he said was "sex oriented" two out of every five times; later Wildmon conceded he might have exaggerated. Indicentally, in their private lives Wildmon and Falwell seemed to have been regular television viewers: Wildmon was reportedly a fan of *Columbo,* and Falwell was dedicated to televised sports.

"What really concerns us," Wildmon once told an interviewer about television, "is the value system being depicted in these programs. Precisely, a value system that says violence is a legitimate way to achieve one's goals in life. A value system that says sex is something to snigger at or what you participate with anyone other than your own spouse." Like many Media Snobs, Wildmon had confused the fantasy theater of television with its antithesis, the real world and the real world's standards for behavior.

The Coalition exerted the most painful pressure thus far on television's advertisers, admonishing them that if they did not end their sponsorship of certain shows, they would be subjected to massive, nationwide boycotts of their products. It is possible of course for such chastising to have reverse, unintended effects: a commercial for Diet Pepsi featuring a blonde sensuously sipping through a straw was credited with a 25 percent increase in sales only after the Moral Majority had pointed a finger at it. This is the pattern of the suggestive Brooke Shields commercials for Calvin Klein blue jeans a year earlier; once the morally indignant began to sputter, sales boomed. Identifying certain shows as salacious or depraved could produce a jump in the ratings and a better market for advertisers.

But the threat of a consumer boycott, especially during fitful economic times, was not to be taken lightly by uneasy advertisers. In June 1981 the president of Procter and Gamble, television's grandest underwriter at some one-half billion dollars annually, announced that the broadcast industry should heed the CBTV, and that his firm would shy away from programs thought to contain gratuitous sex or violence. Even Jerry Falwell was surprised at this show of support, admitting that "until Procter and Gamble's speech, we didn't know so many advertisers agreed with us."

As they had reacted to the earlier challenges, the networks again skirmished with their antagonists. In behalf of CBS, senior vice-president Gene Mater called the Coalition "perhaps the greatest threat to intellectual freedom that we have witnessed in this country in many years." In a public debate with Wildmon, Mater compelled the CBTV leader to say he would disclose his procedures for determining which programs were objectionable and which were not—a pledge Wildmon later reneged on. ABC President James E. Duffy proclaimed, "Make no mistake, what this group seeks to do is to control the content of TV," and went on to say that, according to a network-commissioned poll, just 2 percent of Americans would boycott advertisers. NBC also had a survey conducted of public attitudes; its poll-taker, the Roper Organization, found that only a minute proportion of the population was critical of nine programs the CBTV had listed as offensive.

However, although the networks had put up a good show of resistance, the reformist campaigns of the '70s and the '80s did have a discernible influence upon programming. The 1978 offerings were a departure from previous viewing seasons in that action/adventure shows were little seen on the schedule. Of what the PTA had labeled as the "Ten Worst Shows," all had been cancelled, including *Kojak*. None of these ten shows had been tops in the ratings, but some were on the cusp; for them, the assault by the PTA had an effect. For a time it appeared that the pressure groups were close to sparking Congressional or FCC reviews of programming; the threat was enough to do damage. An anonymous broadcaster quoted in *U.S. News and World Report* said, "We could not afford to risk government interference in programming, up to and including outright censorship. So we pulled back."

Three years later the primary target had shifted from violence to sex. It may not have been for sluggish ratings alone that *Charlie's Angels* and *Soap* were stricken from the 1981 schedule. More popular titillating shows like *Three's Company* and *Dallas* managed to survive, but with fewer innuendoes and more buttons done up. The waitresses in *Making a Living* who wore next to nothing in 1980 returned in 1981 cloaked in respectable outfits. In mid-season a New York advertising executive confided to a *Wall Street Journal* reporter that "the networks have made sure there's not much anyone can object to."

What's wrong with apparently well-intentioned groups marshalling their resources in order to shoulder television content toward what are represented as higher standards? The most profound problem is that, in a democratic land, these groups are promoting the curtailment of the range of broadcast material. They don't simply want to have their voices added to the variety of televised programming, for the medium already permits them ample access to the airwaves. Rather they want to go beyond this, in what amounts to antidemocratic tactics, and restrict the range of television content. In harassing for the elimination of certain themes or certain shows, Media Snobs are acting like censors. Censorship may be tolerated in other nations, but in a democracy it dangerously restricts the range of perceptions and options, and jeopardizes the society's resiliency.

Unfortunately, the censoring of television remains a peril only vaguely appreciated by many Americans. The significance of the free flow of communication, although endorsed by the Constitution, is so abstract, and the consequences of its loss so removed, that the enormity of the issue slips by many. One person who did perceive the dangers and did try to sound the alarm was Fred Silverman. In the last days of his tenure at NBC Silverman addressed the 1981 convention of the American Association of Advertising Agencies, hoping to give heart to the sometimes quavering advertisers: "The Coalition is not asking merely that its views and values be given fair treatment on television, that they compete in the free market of ideas. No. The Coalition wants to restrain trade in ideas and values. It would allow only those they approve of to do business on television. This ideological exclusiveness threatens everyone in this society, not just the networks, not just advertisers."

The program content which reformers want to hack away at has its ultimate source nowhere else than in the free choices of the public-at-large. The fundamental force in American programming, as different from television in many other countries, is the preference of viewers as tabulated and relayed by the ratings services. As we've seen, a highly democratic process is at work here: what the audience indicates it wants will be broadcast. Although there are influences upon programming other than pure popularity—the FCC is one, and the desires of advertisers for viewers of certain demographic characteristics is another—still none is as severely challenging to the democratic propensity of the medium as the pressure from reform groups. In 1977 Robert Blake, the outspoken star of the action/adventure series *Baretta,* reflected about the effect of the cleanup campaigns, "Within a year you're going to have nothing on nighttime television except situation comedies and soap opera," a prediction that in large part came true. Blake complained forcefully, "The people have not spoken. The people speak when they turn the dial—not when a half dozen rich folks who happen to have the time and money go to Washington and bang on desks. Will we give the American Medical Association the right to tell us what should be on TV?"

The strength of the United States has been the strength that emanates when, as Robert Blake put it, the people speak. Our extraordinary success over the last two centuries is due to our grounding whatever decisions have had to be made in the attitudes and opinions of the greater number of citizens. From England we took a judicial system in which crucial judgments were made not by patrician appointees but by groups picked at random to represent a wider population. By enfranchising all adults, government in the United States is also premised on the collective wisdom of the people. Our economic system does not depend on what authorities decree has to be produced, as do many socialist economies, but on the marketplace selections of the public. Not to operate the same way in the case of entertainment fare is tantamount to removing a cornerstone of American life.

What the American people have freely selected for their afterhours diversion are fantasies laden with sex and violence. Other content they have largely spurned, but these myths and fantasies, coming first in one guise and then in another, have proved to be thoroughly therapeutic for weary minds. It is a therapy that virtually everyone needs: one of the findings of the 1981 NBC/Roper poll was that those who declared themselves to be fundamentalists expressed the most concern about televised sex and violence, but then viewed as much television, and the same sorts of programs, as everyone else. If Media Snobs were able to whitewash this chosen programming, they would be doing an immense disservice to viewers. The 55 percent of the viewing audience that turns to *Dallas* every Friday evening would have to turn to something else, because the Rev. Wildmon would have it yanked from the broadcast day (or so he proposed to do). If enough of this censorship went on, the final result would be very much in question, since the noxious psychic energies removed by video fantasies would have to seek their outlet elsewhere.

A great concern is that, should reformers succeed in censoring some television content, they might be able to censor yet more. Newton Minow might discover one night that the action/adventure show he sometimes watches was gone; the Rev. Wildmon might lose his next; Jerry Falwell might see the more vicious of his sports shows removed. Censorship would turn cancerous, moving from entertainment to information, from television to print. The scary spectre of vigilantism spreading and spreading was what convinced Peggy Charren, the head of Action for Children's Television, to turn her back on her fellow reformers: "Perhaps no one will miss the first program forced off the air in the name of morality. But the New Right's censorship crusade will not stop there. What will be the next target? A production of *A Streetcar Named Desire*? A documentary on teenage pregnancy? The news?" Wildmon confessed such expanding ambitions to interviewer Hodding Carter in 1981, saying, "The Lord has more in mind here than television."

It is the sad state of affairs that broadcasters and advertisers, while not capitulating to reform groups, have not been impervious to them either. Many of these executives have allowed their eyes to wander from the goal of audience-determined television. Every concession they made, no matter how minor, was done at the expense of viewers and their needs. Maybe it would embolden broadcasters to recall that if they do more stoutly resist Media Snobs, they can count on the support of the public. A Roper poll in 1979 reported, "The overwhelming opinion of the public is that individual viewers should have the most to say about what is on television, by deciding what they will and will not view." In the 1981 ABC poll, only 2 percent of the population thought that religious organizations should be able to influence programming. Perhaps most Americans would agree with the man from Huron, South Dakota, who wrote to the *Christian Science Monitor,* "Some people or groups are trying to make the television industry conform to their ideas. From there it is a short step to the banning or burning of books, from there a shorter step to complete dictatorship. As for me, I will buy from the firms that are not intimidated by these Hitlerian tactics, and boycott the cowardly ones that give in to this kind of pressure. Thousands of men have fought and died for my right to have any TV program I want, any book I want to read, and my right to go to any church."

The Future

The media future is a fog bank into which we can see only dimly. Swirls and shapes roll about, seem to become distinct, and then retreat. "There's not one person with a definitive idea about where we're going," said Anthony Thomopoulos, president of entertainment at ABC, in 1982. Network personnel like him scan the fog bank most intently, for they are the most vulnerable, entering it first and hugely. What they think they may see there makes them uneasy.

The fog bank is roiled with the turbulence of brand-new delivery systems and their much-touted prospects. Some of these already protrude far from the mists. Cable television wires now extend into a large share of all American households; the proportion grows without letup. Every night millions of cable subscribers can twist their tuners past tens of channels, glancing at an array of offerings that a few years ago was unthinkable. The audiences for religious programs or Spanish broadcasts may be comparatively small, but they are not negligible. Most cable systems also offer for an extra monthly fee a choice of commercial-free movies supplied by such distributors as Home Box Office and Showtime; industry analysts agree that this pay-television service represents the fiercest threat to conventional advertiser-sponsored broadcasting.

The proliferation of cable hookups may be credited in part to another unfolding technology—the stationary satellites that provide economical networking for systems smaller than the Big Three. Twinkling 23,000 miles up in the sky, the satellites with unvarying precision catch the signals beamed up to them and bounce them back to the dish-shaped antennae of cable companies all around the country. It is possible that some day these ground receivers will become so small and inexpensive that every residence will be able to pick up satellite transmissions directly, and the broadcasting systems we now have will wither away. Network executives cannot relish the thought of that.

From a network point of view the most invidious technology lurking in the fog bank of the future are video-cassettes and their cousins, video-discs. Horrors: the programming recorded on these devices can be played at the convenience of the viewer, weeks and years later, just as records and audio-cassettes can be now. If these technologies catch on widely, they will free the audience from the tyrannies of the broadcast schedule, and will spell the end of the timely advertising that currently sustains programming and broadcasters. The video world would be turned on its ear.

The prospering of these alternative delivery systems owes much to the benevolence of the Federal Communications Commission, which has been reworking regulatory codes to suit the upstart technologies. In decades past the FCC appeared to favor the three large networks at the expense of any competitors; with hindsight we can see that there may have been reason to try to stabilize what was a large, important, but initially amorphous enterprise. As the television industry has matured, however, the FCC has made more decisions that have encouraged newer modes of transmission. The Commissioners' unconcealed hope is to make the communications environment less oligarchic, more competitive, and more responsive to the nation's best interests, as seen from their vantage point.

Taking stock of the ever more visible variety in delivery systems, some futurists and television critics prophesy programming of widening range and ascending quality. Alvin Toffler wrote in his *The Third Wave,* "We're going to move from a few images distributed widely to many images distributed narrowly." This "narrowcasting," as it has been dubbed, is supposed to bring Americans a rich collection of tasteful, imaginative programs. The golden age of civilized video is right before us, these futurists claim.

What is more distinct, less disputable, about the hazy remaining years of the 20th century is the aging of the audience. The maturing baby-boom generation is so large, and the birthrate has remained so low from 1965 on, that the average age of Americans will continue to step upward, past 30 in 1980, past 33 in 1990.

This aging of the audience brings with it several changes that portend to reshape television-viewing in the 1980s and '90s. Middle-aged people are more sedentary; from now through the end of the century, people are likely to be staying home more, and so will be even more receptive to the medium. Americans in mid-career should have more buying power and may not hallow "free" television as much as the public did in previous years. And a maturing audience is bound to have maturing taste in content.

Other attributes of the future audience can be deduced not from generalizations about the middle-aged but from the unique qualities of Americans at this point in our history. We are becoming less preoccupied with family life, to judge from the low birthrates and the shrinking size of the average household, as related in Census data. It has been calculated that by 1990 two-thirds of American households will contain no children. The segment of the population that lived alone doubled during the 1970s and will in all likelihood double again in the 1980s. In the years ahead, therefore, attention will center not so much on childrearing and kindred activities as on adult careers and living styles. The steady rise in full-time employment for women shows no signs of abating, so women as well as men will be hitting full stride in their lives' work then. Greater application during the day, fewer distractions in the evening—it adds up to more intense prime-time viewing than was possible in the households of the '60s and '70s.

What kind of programming are these mature, hard-working individuals going to want at day's end? They will seek out the shows that help them repair from their labors; even more than now, they will prize fantasies. The fantasies they will demand, though, will have to be a bit more gripping and satisfying than much of what is presently offered on television. The slight, frothy situation comedies with stars in their twenties are going to be less visible on screens, while along will come slightly older characters who are up to more poignant, full-blooded activity. We can be certain that sex and violence will be as endemic as ever if the sanctimonies of Media Snobs are deflected. Of all forms of video entertainment, it is movies that best meet this requirement for engrossing, full-blown fantasy.

And of all ways of getting movies onto home sets, it is pay-cable systems that seem best situated to respond to the quickening demand. They can deliver recent, uncut movies that are unsullied by commercial minutes, and thus they can best gratify viewers' heightened fantasy desires. Future viewers, wanting their Hollywood dream material straight and unadulterated, will be in a position to pay whatever it costs for the service. By 1990 the chances are good that pay-cable, rather than movie theaters, will be the chief form of release for new films. This switch will be deadly for cinema owners, but a boon to movie producers who recognize that the ranks of their young theater audience (in 1981 70 percent of all tickets were bought by people under the age of 30) are soon to be demographically depleted. It is only logical that these producers will accelerate the changeover by creating more movies for the older, stay-at-home cable audience. In upcoming films we can expect fewer extraterrestrials and more ordinary people.

In short, it appears that the network monopoly of the televised fantasy business is doomed. The networks are simply unable to provide the interruption-free fantasies that Americans are coming to crave. When the networks ought to be working in their own self-interest to reduce prime-time commercial clutter, they are not. Instead their response to shrinking audiences and revenues is to tolerate yet more commercials, and provoke yet more audience disenchantment. In their frenzy to halt the viewer defection the networks have been juggling the schedule wildly—during one sample month in the 1981-82 viewing season, only 23 of 53 serials ran in their regular time slots. The sense of ease that viewers seek after their workdays is undermined by such manic maneuvering. In 1982 Ogilvy and Mather advertising agency felt safe in predicting to its clients that the network share of the prime-time audience, which had been almost 90 percent in 1980, would slip below 60 percent by 1990.

Some forms of network fantasy material will probably weather the decline better than others. A broadcast schedule where shows begin and end at the hour or half-hour mark is needed to coordinate networks and affiliates, but it works against the kinds of programs like movies that are better governed by the time requirements of their own internal dramatic movement; however, such schedule practices do not hurt every type of video fantasy. Situation comedies, for example, fit well into half-hour frames. And given their convulsive nature, situation comedies are not seriously marred by commercial intrusions. Although they will be playing to smaller audiences, situation comedies will be seen on network television for years to come.

Sports broadcasts, as a sort of fantasy material, also seem unblemished by commercial interruptions. The "breaks in the action," when baseball teams change sides or football teams regroup or take a time out, are tolerable occasions for a message from a sponsor, especially if it is diverting in its own right. Viewer loyalty for sponsored games may not weaken, and the networks may be able to continue outbidding cable challengers for broadcast rights. To the extent that sports are televised during the daytime, the acceptance of the interspersed commercial messages may remain high, for it is during non-prime-time hours that viewers are least offended by the quasi-informative content of advertisements.

This is part of the reason that daytime programming from the networks is little threatened. Those who view during the day tend to be those who do most of the household purchasing, so information about what is being offered in the marketplace is not as irritating to them as it is to those who view only after a day's work. The networks appear to have a firm grip on the economics and production of soap operas, and the loss of these shows to cable is not soon anticipated.

The kind of programming least likely to slip away from the networks in the decades ahead is the kind many feel they execute the best—news. The issue of commercial interruptions may be least galling here, for viewers watch the news in the most instructible frame of mind they ever bring to television; they can more easily assent to advertising information during a news broadcast than during their spell of video fantasy. Since people's interest in news is known to expand as they age, it is safe to predict that the appetite of the American audience for real-world information is going to grow in the future. Affiliates will soon come to concede that, in the face of a dwindling audience for their fantasy programming, an expansion of news broadcasts makes good sense.

In sum, the viewers of the future will be watching more television but less network broadcasting. It is the need for fantasy material, the main force behind all media participation, that is creating these changes. As decreasing family obligations grant more opportunity for closer viewing, Americans will surrender the familiarity of serials for the more intense experience of movies. And they will cheerfully pay for the privilege. Movies brought into our homes by cable systems will become our well-purchased bedtime stories.

Although our total time with the medium will rise, proportionally we will take our video content much as we always have. That is, about 75 percent of our viewing hours will still be devoted to what Gerhardt Wiebe called *restorative* content—the fantasies coming increasingly as cabled movies, decreasingly as network shows. About 15 percent of what Americans watch will be material of the *maintenance* sort—largely the soap operas that will remain in network hands. And as before we'll spend 10 percent of the expanded television time with the news and other reality material which Wiebe labeled *directive*.

Let's close by returning to Russell Baker's *New York Times* column on the Ronnie Zamora trial. Baker had ridiculed the defense presented by the lawyer who argued that television had made Ronnie commit murder. According to Baker, if television violence could actually instigate real-world violence, then why didn't television portrayals of people happily at work produce happy workers in real life? Baker didn't have to pick the topic of work for his analogy revealing the uninstructive nature of television fantasy. He might have picked friendship, and asked if people in situation comedies are so friendly, then why aren't real Americans? Or asked if television is so full of laughs, then why aren't our lives? But he selected work because he wanted to go on and make an observation about the relationship of television, hostility, and work. He ended his column saying that television doesn't direct our behavior: "Television's supposedly terrible power to control our lives crumbles before our iron distaste for the daily grind. Perhaps its power to turn us to violence is also overrated. Maybe it is not television at all that does this to people. It might just as sensibly be blamed on the daily grind."

Baker need not have been so tentative. It is indeed the daily grind that can make us uneasy and resentful, even antagonistic. Earning a living or maintaining a household brings us pleasures and benefits, to be sure, but there are more than a few costs. It is televised fantasies that help make up for those costs. Whenever television is blamed for our failings, a criminal disservice and misunderstanding is being perpetuated; the cure is being treated as the curse.

Television heals, not harms. This is the reason that almost all of us, despite whatever attitudes we profess to hold, resort to the medium; our deep-lying, health-seeking needs for psychological discharge and renewal lead us to embrace it. Americans' extensive use of television is not a comment on personal frailties, but on strengths. The better use we make of televised fantasies, the more prodigious our future will be.

SOURCES

Much of the research for this book was done at the library of the Television Information Office in New York City. My great appreciation goes to head librarian James Poteat and his staff; they were unfailingly helpful and gracious.

Listed below are the more important sources of the material cited in this book.

Adler, Richard, *et al.,* eds. *Research on the Effects of Television Advertising on Children.* Washington: National Science Foundation, 1977.

Allen, Charles L. "Photographing the TV Audience." *Journal of Advertising Research* 5 (March 1965): 2–8.

Altheide, David L. *Creating Reality: How TV News Distorts Events.* Beverly Hills, California: Sage, 1974.

Andrews, Bart. *Lucy and Ricky and Fred and Ethel: The Story of "I Love Lucy."* New York: Dutton, 1976.

Atwan, Robert; McQuade, Donald; and Wright, John W. *Edsels, Luckies, and Frigidaires: Advertising the American Way.* New York: Dell, 1979.

Baker, Russell. "As the World Turns." *The New York Times Magazine,* October 30, 1977, p. 12.

Baker, Russell. "Heehaw." *The New York Times Magazine,* December 4, 1977, p. 16.

Bandura, Albert, et al. "Transmission of Aggression Through Imitation of Aggressive Models." *Journal of Abnormal and Social Psychology* 63 (1961): 575–82.

Barber, James David. "Not *The New York Times*: What Network News Should Be." *The Washington Monthly,* September 1979, pp. 14–21.

Barnard, Charles. "An Oriental Mystery." *TV Guide,* January 28, 1978, pp. 2–8.

Barnouw, Erik. *The Sponsor.* New York: Oxford, 1978.

Barnouw, Erik. *The Tube of Plenty: The Evolution of American Television.* New York: Oxford, 1975.

Bauer, Raymond A. "Communication as Transaction." In *The Obstinate Audience,* edited by Donald E. Payne, pp. 3–12. Ann Arbor, Michigan: Foundation for Research on Human Behavior, 1965.

Bauer, Raymond A. "The Obstinate Audience: The Influence Process from the Point of View of Social Communication." In *The Process and Effects of Mass Communication,* edited by Wilbur Schramm and Donald F. Brooks, pp. 326–46. Urbana: University of Illinois Press, 1971.

Bauer, Raymond A. and Greyser, Stephen A. *Advertising in America: The Consumer View.* Boston: Division of Research, Graduate School of Business Administration, Harvard University, 1968.

Bechtel, Robert B., *et al.* "Correlates Between Observed Behavior and Questionnaire Responses on Television Viewing." In *Television and Social Behavior,* The Surgeon General's Scientific Advisory Committee, Volume IV: Television in Day-to-Day Life, pp. 274–345. Washington: U.S. Government Printing Office, 1972.

Bettelheim, Bruno. *The Uses of Enchantment: The Meaning and Importance of Fairy Tales.* New York: Knopf, 1976.

Blumstein, Alfred, *et al.* "Crime, Punishment, and Demographics." *American Demographics,* October 1980, pp. 32–37.

Bogart, Leo. *The Age of Television.* New York: Ungar, 1972.

Bower, Robert T. *Television and the Public.* New York: Holt, 1973.

Brown, Les. *Television: The Business Behind the Box.* New York: Harcourt, 1971.

Carpenter, Edmund. *Oh, What a Blow That Phantom Gave Me!* New York: Holt, 1972.

Comstock, George, *et al. Television and Human Behavior.* New York: Columbia University Press, 1978.

Cook, Thomas D., *et al. "Sesame Street" Revisited.* New York: Russell Sage Foundation, 1975.

Diamond, Edwin. *The Tin Kazoo: Television, Politics and the News.* Cambridge, Massachusetts: MIT Press, 1975.

Edmondson, Madeleine, and Rounds, David. *The Soaps: Daytime Serials of Radio and TV.* New York: Stein and Day, 1973.

Efron, Edith. "Does TV Violence Affect Our Society? No." *TV Guide,* June 14, 1975, pp. 22–28.

Efron, Edith. "A Million Dollar Misunderstanding." *TV Guide,* November 11, 1972, November 18, 1972, and November 25, 1972.

Epstein, Edward Jay. *News from Nowhere: Television and the News.* New York: Random House, 1973.

Ferdinand, Theodore N. "The Criminal Patterns of Boston Since 1849." *American Journal of Sociology* 73 (1967): 84–99.

Feshbach, Seymour. "The Drive-Reducing Function of Fantasy Behavior." *Journal of Abnormal and Social Psychology* 50 (1955): 3–11.

Feshbach, Seymour. "Reality and Fantasy in Filmed Violence." In *Television and Social Behavior,* The Surgeon General's Scientific Advisory Committee, Volume II: Television and Social Learning, pp. 318–45. Washington: U.S. Government Printing Office, 1972.

Feshbach, Seymour. "The Role of Fantasy in the Response to Television." *Journal of Social Issues* 32 (1976): 71–85.

Feshbach, Seymour. "The Stimulating Versus Cathartic Effects of a Vicarious Aggressive Activity." *Journal of Abnormal and Social Psychology* 63 (1961): 381–85.

Feshbach, Seymour, and Singer, Robert D. *Television and Aggression.* San Francisco: Jossey-Bass, 1971.

Finkelstein, Sidney. *Sense and Nonsense in McLuhan.* New York: International Publishers, 1969.

Fowles, Barbara R. "A Child and His Television Set: What Is the Nature of the Relationship?" *Education and Urban Society* 10 (1977): 89–102.

Freud, Sigmund. *Jokes and Their Relation to the Unconscious.* Translated by James Strachey. New York: Norton, 1960.

Freud, Sigmund. "The Relation of the Poet to Day-Dreaming." In *Collected Papers,* edited by Ernest Jones, translated by Joan Riviere, Volume 4, pp. 173–83. New York: Basic Books, 1959.

Gans, Herbert J. *Deciding What's News.* New York; Pantheon, 1979.

Gans, Herbert J. *Popular Culture and High Culture.* New York: Basic Books, 1974.

Gans, Herbert J. *The Uses of Television and Their Educational Implications.* New York: Center for Urban Education, 1968.

Gerbner, George. "Cultural Indicators: The Case of Violence in Television Drama." *The Annals of the American Academy of Political and Social Science.* Volume 388 (1970): 69–81.

Gerbner, George, and Gross, Larry. "The Scary World of TV's Heavy Viewer." *Psychology Today,* April 1976, pp. 41–45, 89.

Goldsen, Rose. *The Show and Tell Machine.* New York: Dial, 1975.

Goldsen, Rose. "Throwaway Husbands, Wives, and Lovers." *Human Behavior,* December 1975, pp. 64–69.

Graham, Fred P. "A Contemporary History of American Crime." In *The History of Violence in America,* edited by Hugh Davis Graham and Ted Robert Gurr, pp. 485–504. New York: Bantam, 1969.

Granzberg, Gary, *et al.* "New Magic for Old TV in Cree Culture." *Journal of Communication* 27 (1977): 154–57.

Greenfield, Jeff. *Television: The First Fifty Years.* New York: Abrams, 1977.

Hapkiewicz, Walter G. "Children's Reaction to Cartoon Violence." *Journal of Clinical Child Psychology* 8 (1979): 30–34.

Hazard, William R. "Anxiety and Preference for Television Fantasy." *Journalism Quarterly* 44 (1967): 461–69.

Henniger, Daniel. "A Meditation on the Joy of TV Commercials." *Wall Street Journal,* October 20, 1980, p. 22.

Henry, William A., III. "News as Entertainment: The Search for Dramatic Unity." In *What's News: The Media in American Society,* edited by Elie Abel, pp. 133–58. San Francisco: Institute for Contemporary Studies, 1981.

Hirsch, Paul. "The Scary World of the Nonviewer and Other Anomalies —A Reanalysis of Gerbner *et al.*'s Findings on Cultivation Analysis. Part I." *Communication Research* 7 (1980): 403–45. "Part II." 8 (1981): 3–37.

Howitt, Dennis, and Cumberbatch, Guy. *Mass Media Violence and Society.* New York: Wiley, 1975.

Johnson, Nicholas. *How to Talk Back to Your Television Set.* Boston: Little, Brown, 1970.

Jones, Landon. *Great Expectations: America and the Baby Boom Generation.* New York: Coward, McCann, and Geoghegan, 1980.

Kaplan, R. M., and Singer, R. D. "Television Violence and Viewer Aggression: A Re-examination of the Evidence." *Journal of Social Issues* 32 (1976): 35–70.

Katzman, Natan. "Television Soap Operas: What's Been Going on Anyway?" *Public Opinion Quarterly* 36 (1972): 200–12.

Kilguss, Anne F. "Using Soap Operas as a Therapeutic Tool." *Social Casework* 55 (1974): 525.

Klapper, Joseph. *The Effects of Mass Communication.* New York: Free Press, 1960.

Kuhns, William. *Why We Watch Them: Interpreting TV Shows.* New York: Benzinger, 1970.

Lang, Kurt, and Lang, Gladys Engel. *Politics and Television.* Chicago: Quadrangle, 1968.

Levy, Mark Robert. "The Uses-and-Gratifications of Television News." Ph.D. dissertation, Columbia University, 1977.

Lopiparo, Jerome J. "Aggression on TV Could Be Helping Our Children." *Intellect,* April 1977, pp. 345–46.

Lyle, Jack, and Hoffman, Heidi R. "Children's Use of Television and Other Media." In *Television and Social Behavior,* The Surgeon General's Scientific Advisory Committee, Volume IV: Television in Day-to-Day Life, pp. 129–256. Washington: U.S. Government Printing Office, 1972.

Mander, Jerry. *Four Arguments for the Elimination of Television.* New York: Morrow, 1978.

Mankiewicz, Frank, and Swerdlow, Joel. *Remote Control: Television and the Manipulation of American Life.* New York: Ballantine, 1978.

Mayer, Martin. *About Television.* New York: Harper and Row, 1972.

McCarney, Hugh R. "The Television Rating System: A Descriptive Study of its Development and Implications." Ph.D. dissertation, New York University, 1978.

McGinnis, Joe. *The Selling of the President 1968.* New York: Trident, 1970.

McIntyre, Jennie J.; Teevan, James J. Jr.; and Hartnagel, Timothy F. "Television Violence and Deviant Behavior." In *Television and Social Behavior,* The Surgeon General's Scientific Advisory Committee, Volume IV: Television in Day-to-Day Life, pp. 383–435. Washington: U.S. Government Printing Office, 1972.
Reanalysis and revised findings published as: Hartnagel, Timothy F.; Teevan, James J. Jr.; and McIntyre, Jennie J. "Television Violence and Violent Behavior." *Social Forces* 54 (1975): 341–49.

McLuhan, Marshall. *Understanding Media: The Extensions of Man.* New York: McGraw-Hill, 1964.

Milgram, Stanley, and Shotland, R. Lance. *Television and Antisocial Behavior: Field Experiments.* New York: Academic Press, 1973.

Mindak, William H., and Hursh, Gerald D. "Television's Functions on the Assassination Weekend." In *The Kennedy Assassination and the American Public,* edited by Bradley S. Greenberg and Edwin B. Parker, pp. 130–41. Stanford, California: Stanford University Press, 1965.

Nelson, Phillip. "Advertising as Information Once More." In *Issues in Advertising,* edited by David G. Tuerck, pp. 133–60. Washington: American Enterprise Institute for Public Policy Research, 1978.

Nelson, Phillip. "The Economic Values of Advertising." In *Advertising and Society,* edited by Yale Brozen, pp. 43–66. New York: New York University Press, 1974.

Newcomb, Horace. *TV: The Most Popular Art.* Garden City, New York: Anchor, 1974.

Patterson, Thomas E. *The Mass Media Election: How Americans Choose Their President.* New York: Praeger, 1980.

Patterson, Thomas E., and McClure, Robert D. *The Unseeing Eye: The Myth of Television Power in National Elections.* New York: Putnam's, 1976.

Pearlin, Leonard I. "Social and Personal Stress and Escape Television Viewing." *Public Opinion Quarterly* 23 (1959): 255–59.

Price, Jonathan. *The Best Thing on TV: Commercials.* New York: Penguin, 1978.

Robinson, John P. "Television and Leisure Time: A New Scenario." *Journal of Communication* 31 (1981): 120–30.

Robinson, John P. "Toward Defining the Functions of Television." In *Television and Social Behavior,* The Surgeon General's Scientific Advisory Committee, Volume IV: Television in Day-to-Day Life, pp. 568–603. Washington: U.S. Government Printing Office, 1972.

Roper Organization. *Public Perceptions of Television and Other Mass Media: A Twenty Year Review 1959–1978.* New York: Television Information Office, 1979.

Schramm, Wilbur, *et al. Television In the Lives of Our Children.* Stanford, California: Stanford University Press, 1961.

Seldes, Gilbert. *The Public Arts.* New York: Simon and Schuster, 1956.

Shanks, Bob. *The Cool Fire: How to Make It in Television.* New York: Vintage, 1976.

Silverman, Fred. "An Analysis of ABC Television Network Programming From February 1953 to October 1959." Master's thesis, Ohio State University, 1959.

Steiner, Gary A. *The People Look at Television.* New York: Knopf, 1963.

The Surgeon General's Scientific Advisory Committee on Television and Social Behavior. *Television and Social Behavior.* Washington: U.S. Government Printing Office, 1972.

Tan, Alexis S. "Why TV Is Missed: A Functional Analysis." *Journal of Broadcasting* 21 (1977): 371–80.

Thomas, Margaret Hanratty, and Drabman, R. S. "Some New Faces of the One-Eyed Monster." Paper presented at the meeting of the Society for Research in Child Development, Denver, 1974.

Thomas, Margaret Hanratty, *et al.* "Desensitization to Portrayals of Real Life Aggression as a Function of Exposure to Television Violence." *Journal of Personality and Social Psychology* 35 (1977): 450–58.

Trost, Cathy, and Grzech, Ellen. "Could You Kick the TV Habit?" *Detroit Free Press.* Seven articles in September and October 1977.

Tuchman, Gaye. *Making News: A Study in the Construction of Reality.* New York: Free Press, 1978.

Tyler, Ralph. "Learning in America." *The Center Magazine,* November-December 1977, pp. 50–54.

Wheeler, Michael. *Lies, Damn Lies, and Statistics.* New York: Dell, 1976.

Whiteside, Thomas. "Din-din." *The New Yorker,* November 1, 1976, p. 51.

Wiebe, Gerhardt D. "Two Psychological Factors in Audience Behavior." *Public Opinion Quarterly* 33 (1970): 523–36.

Winick, Mariann Pezzella, and Winick, Charles. *The Television Experience: What Children See.* Beverly Hills, California: Sage, 1979.

Winn, Marie. *The Plug-In Drug: Television, Children, and the Family.* New York: Viking, 1976.

Index